Making Meanings in Mathematics

A collection of extended and refereed papers from BSRLM – the British Society for Research into Learning Mathematics

Edited by
Laurinda Brown

Visions of Mathematics No.2

Advances in Mathematics Education No.1

1-85853-102-0
ISSN: 1463-6441

ADVANCES IN MATHEMATICS EDUCATION

This book is produced in collaboration between QED and the British Society for Research into Mathematics Learning (BSRLM). It is the second in QED's series "Visions of Mathematics", and the *first* in the new series *Advances in Mathematics Education.*

The aim of this new series is to present informative and accessible accounts of the work of the British Society for Research into Mathematics Learning (BSRLM) to the widest possible audience. Each volume will contain papers developed from presentations at the regular conferences of the Society. The quality of each paper has been assured by the process of peer review.

The theme *Making Meanings in Mathematics* runs through the contributions to this first collection. The authors introduce and exploit multiple perspectives on what it means to learn mathematics at various levels. They demonstrate how, from its roots in psychology, Mathematics Education has matured and expanded to embrace 'lenses' and frameworks from sociology, philosophy, semiotics, hermeneutics and post-structuralism. The papers reflect the range of methods deployed – empirical, statistical, theoretical – to illuminate mathematical meaning-making.

This volume contains research-based insights into:

- the influence of new technologies on mathematical thinking
- students' difficulties in progressing to advanced study of mathematics
- the interplay between examples and generalisations
- issues affecting undergraduate students' mathematical learning
- social justice and mathematics education
- the role of language and other systems of signs (semiotics)
- the linguistic framing of mathematical thought.

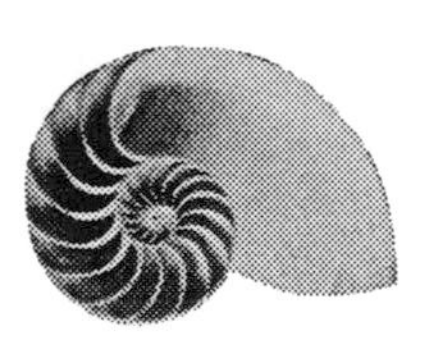

CONTENTS

BSRLM OFFICERS AND CONTRIBUTORS

http://www.warwick.ac.uk/wie/merc/bsrlm/

"The BSRLM ... acts as a major forum for research in mathematics education ... It is both an environment for supporting new researchers and a forum for established ones"
(from Introduction by Laurinda Brown)

Chair
Paul Ernest, University of Exeter, School of Education, Exeter EX1 2LU
Tel: 01392-264857 Fax: 01392-264736
p.ernest@exeter.ac.uk

Secretary
Candia Morgan, Mathematical Sciences, Institute of Education, University of London, 20 Bedford Way, London WC1H 0AL. Tel: 0171 612 6677 Fax: 0171 612 6686
temscrm@ioe.ac.uk

Membership
Tim Rowland, Institute of Education, University of London, 20 Bedford Way, London WC1H 0AL. Tel: 0171 612 6231 Fax: 0171 612 6230
t.rowland@ioe.ac.uk

Treasurer
Keith Jones, School of Education, University of Southampton, Southampton SO17 1BJ.
Tel: 01703 592449 Fax: 01703 593556
d.k.jones@soton.ac.uk

Contributors

Liz Bills, The University of Warwick, Institute of Education, Coventry CV4 7AL
liz.bills@warwick.ac.uk

Tony Brown, Didsbury School of Education, Manchester Metropolitan University, 799 Wilmslow Road, Didsbury, Manchester M20 2RR
A.M.Brown@mmu.ac.uk

Tony Cotton, Centre for the Study of Mathematics Education, University of Nottingham, University Park, Nottingham, NG7 2RD
Tony.Cotton@nottingham.ac.uk

Brian Hudson, Mathematics Education Research Group, School of Education, Sheffield Hallam University, Collegiate Crescent Campus, Sheffield S10 2BP
B.G.Hudson@shu.ac.uk

Elena Nardi, School of Education, University of East Anglia, Norwich NR4 7TJ
e.nardi@uea.ac.uk

Declan O'Reilly, The University of Sheffield, Division of Education, The Education Building, 388 Glossop Road, Sheffield S10 2JA
D.OReilly@sheffield.ac.uk

Adam Vile, School of Computing, Information Systems and Mathematics, South Bank University, London SE1 0AA
vileawa@sbu.ac.uk

Dylan Wiliam, King's College, University of London, Cornwall House, Waterloo Road, London SE1 8WA
dylan.wiliam@kcl.ac.uk

INTRODUCTION

Laurinda Brown

University of Bristol

'BSRLM is an organisation which acts as a major forum for research in mathematics education in the United Kingdom. It is both an environment for supporting new researchers and a forum for established ones.'

At the end of this year, I complete my term of office as Treasurer for the British Society for Research into Learning Mathematics. Editing this first collection of edited and reviewed papers arising from BSRLM conferences presents me with an opportunity to reflect on the work of the Society.

When I think about BSRLM, my images are mainly of the day conferences where the format is of the presentation of research papers which can be work in progress or the dissemination of completed pieces of work. I first attended conferences of the then BSPLM (the P standing for Psychology) when I was working in a 'Teacher as Researcher' group facilitated by Joan Yates in the late 70s and early 80s. I was a practising teacher in a school in the Bristol area and the first conference I attended was at the University of Bath, geographically quite local to Bristol. I can remember the excitement with which I talked with people who had been simply names on papers and heard them presenting their new ideas. Recently a group of my M.Ed. students attended the day conference in Bristol and their reactions seemed similar (written up in the June 1998 issue of *Mathematics Teaching*). BSRLM is not a Society for researchers who have nothing to say to teachers, but is open to 'anyone interested in the area of mathematics education' and, of course, many teachers are also researchers and many researchers have been practising teachers.

For each of the conferences of the Society a non-edited, non-reviewed copy of the proceedings is produced. These proceedings consist of short papers and their aim is 'to communicate to the research community the collective research represented as quickly as possible'. There is the hope that members use these proceedings to give feedback to authors and that through discussion and debate an energetic and critical research community is sustained. Given the quality of discussion and debate which exists within the community a more substantial publication was envisaged, whilst

Steve Lerman was Chair of the Society, and this volume of what we intend to be an annual collection of papers is the outcome.

In editing this collection of papers I am again excited by the ideas represented and therefore discussed within the community of researchers and teachers which is BSRLM. This is not a closed, inward-looking community but there are influences from social practice theory (Hudson *et al.* p. 1), semiotics (Vile, p. 87) and Derrida's work in the study of language (Brown, p. 11). A mix here of established researchers and of studies which contributed towards the award of a PhD. It might be asked what the use of these theoretical perspectives might be for practising teachers but there is a strong strand of finding ways of describing 'meaning-making' throughout these papers and it is this process which is at the heart of learning mathematics, be it the 'novice mathematician' undergraduate (Nardi, p. 57), young children working with the computer environment 'Boxer' (O'Reilly, p. 23), through trying to describe the link between examples and generalisation (Bills and Rowland, p. 103) or through trying to find explanations for the seeming difficulties students might experience in the transition from GCSE to A-level mathematics (Wiliam *et al.*, p. 41).

Thanks must go in good measure to the authors of these papers for providing me with disks to work with after a long and illuminating review process. The responsibility for any errors which have slipped through the net reside with me. One view of the editor's task is to make a style of diversity so that there is no jolt from one paper to the next and the reader can concentrate on the text. For this reason my last word resides with a paper which quotes Lyotard (1992): *Reading is an exercise in listening* and gives us the space in which to do so (Cotton, p. 69) whilst discussing 'social justice and the mathematics classroom'. It is possible, having thought a lot about this whilst working with the details of the text, that such conformity of text becomes a barrier to the meaning. Whatever, I hope that you, the reader, find this collection of papers as stimulating to read as I did to edit.

Laurinda Brown

July 1998

AN ANALYSIS OF STUDENT TALK IN 'RE-LEARNING' ALGEBRA: FROM INDIVIDUAL COGNITION TO SOCIAL PRACTICE

Brian Hudson, Susan Elliott and Sylvia Johnson

Sheffield Hallam University

Abstract

In this paper we report on a study with the aim of investigating how a focus on language and meaning can assist students in reconstructing algebraic knowledge. The project is set in the context of ongoing work with students in Higher Education who need to develop their understanding of algebra if they are to make substantial progress within their undergraduate studies. The project is based upon a belief that students' difficulties with algebra are language-related. We have collected extensive data by means of video taped sessions involving the students talking about their own understandings of algebra. The students involved were drawn from courses in initial teacher education and engineering. This paper presents a detailed analysis of the responses of one student and discusses the ways in which this shifted our attention as researchers from looking at our data from the perspective of individual cognition towards one informed by social practice theory.

Introduction

The Re-Learning Algebra project grew out of the difficulties many students have with algebra which have been observed in the course of working in the arena of Academic Mathematics Support at Sheffield Hallam University. These students have considerable prior experience with algebra and many have undergone years of drill and practice. They have encountered algebra as both an abstract topic in its own right and also within various contexts. Therefore any additional help offered to such students clearly needed to take account of previous experience but also needed to have a different emphasis. An approach which was seen to be successful in practice involved encouraging interaction using group activities in which the students could share their understanding and experience. The activities also addressed the use and development of algebraic language and have previously been reported in Elliott and Johnson (1995).

The three researchers involved in this project brought differing initial perspectives and emphases, including:

- an interest in the use of vocabulary and terminology, such as *solve* and *simplify*
- David Pimm's (1995) discussion of the notion of a mathematical register
- an emphasis on the language competence underpinning the development of symbolic representation
- the Vygotskian (1962) emphasis on the social and communicative aspects of language and on speech as an instrument of thought itself, that is, as a psychological tool.

We were all agreed, however, that the shared understanding of meanings is crucial both for our students re-learning algebra and for ourselves as researchers.

Theoretical framework

Given the initial aim of this study, which was to investigate how a focus on language and meaning can assist students in reconstructing algebraic knowledge, we have sought to develop a theoretical framework which takes account of this emphasis on language and meaning. As indicated, each of the researchers involved in this study brought particular emphases, although we were unanimous in the belief that students' talking about their own understandings of algebra and the sharing of meanings was crucial for their re-learning of algebra. In relation to this our thinking has developed through our discussions of the related literature and also through our analysis of the data.

A key influence has been the work of Lev Vygotsky (1962), underpinning which is a central assumption that socio-cultural factors are essential in the development of mind. Intellectual development is seen in terms of meaning making, memory, attention, thinking, perception and consciousness which evolves from the interpersonal to the intrapersonal. The social dimension is primary in time and fact - the individual dimension is derivative and secondary. In discussing the influence of such a perspective, Stephen Lerman (1996) describes language as providing the tools for thought, and carrying the cultural inheritance of the communities (ethnic, gender, class etc.) in which the individual grows up. He argues that language is not seen as giving structure to the already conscious cognising mind; rather the mind is constituted in discursive practices. Thus the semiotic function becomes the focus of study, rather than the mental structures (for Piaget). He also draws on Kozulin (1990) in order to help distinguish the links between thought, language and consciousness and hence sense and meaning. Kozulin argues that it is incorrect to consider language as correlative of thought but that rather language is a correlative of consciousness. He argues that the mode of language correlative to consciousness is meanings and that the work of consciousness with meanings leads to the generation of sense, and in the process consciousness acquires a sensible (meaningful) structure. He proposes that to study human consciousness means to study this sensible structure, and that verbal meaning is the methodological unit of this study. He also proposes that at the concrete level we should be concerned with the specific "sense generating" activity that changes the consciousness of a person.

In exploring the notions of sense and meaning further some useful ideas were drawn from the field of activity theory. In particular Erik Schultz (1994) offers some interpretations of sense and meaning when writing about the hermeneutical aspects of activity theory, which we found to be helpful in interpreting some of our data. He proposes that the purpose or intention of a cultural product is the meaning and further that meaning is a kind of 'cultural intention' in a supra-individual fashion. In all cultural products there is an intention to be found, and in finding it, we interpret the meaning of the product. Sense is the interpretation one makes of the meaning. He argues further that language is a special kind of cultural product. We also found the work on activity theory of Kathryn Crawford (1996) to be relevant. She highlights how activity denotes personal (or group) involvement, intent and commitment that is not reflected in the usual meanings of the word in English. She draws attention to the fact that Vygotsky wrote about activity in general terms to describe the personal and voluntary engagement of people in context - the ways in which they subjectively perceive their needs and the possibilities of a situation and choose actions to reach personally meaningful goals. In building upon Vygotsky's work, Leont'ev, Davydov and others made clear distinctions between conscious actions and relatively unconscious and automated operations. Operations are seen as habits and automated procedures that are carried without conscious intellectual effort. In reporting upon her own research Crawford reports that 'traditional forms of instruction' were effective in establishing student abilities to implement standard mathematical techniques - or operations. However students in the study showed no spontaneous disposition to engage personally with the mathematical task, i.e. activity. She reports that her students were inexperienced and unskilled in interpreting the meaning of mathematical information, defining a problem, selecting strategies and evaluating the results of problem solving efforts. That is they were inexperienced in mathematical actions and that many lacked the conceptual framework to 'make sense of' non-standard problems. She also reports a study of first year undergraduate students which found that more than 80% of the sample studied viewed mathematics as a set of rote learned rules and techniques and approached mathematics learning in a fragmented fashion with the intent to reproduce using paper and pencil, axioms and standard techniques for examinations. Even the most successful students of mathematics, at the end of their school experience, viewed mathematics as a series of operations - techniques and rules to be implemented. These findings resonated with our own observations prior to embarking on this particular study.

What we did

Data were collected by means of the video recording of a series of one-hour sessions with four groups of students during March 1995. The students were drawn from one of two secondary BEd courses (a two-year and a four-year honours degree route) and also from a BSc and an HND course in Engineering. The groups had two or three sessions each.

A series of tasks was devised, based upon the practice in the Maths Support sessions, which were designed to get the students talking together about their understanding of algebra. For example the first activity which was carried out with each group involved 'Algebraic Pairs'. In this activity each pair of students is given a set of cards with a pair of algebraic expressions on each. The task is to decide if the two expressions are always, sometimes or never equal. Another activity was to ask them to explain what they understood by mathematical words such as expression, equation, function, variable etc. The sessions were carried out in a small TV studio on the university campus and the videotapes were subsequently transcribed.

The initial data analysis involved the three researchers viewing the videotapes collaboratively and discussing reactions and questions arising. Following the tape transcription this process was repeated with the transcripts both as a group and also on an individual and paired basis. There then followed in-depth and detailed analyses of particular sections of the videotape transcripts involving the identification of emergent themes based upon particular forms of language use. We also held two internal university research seminars during this period.

Our aim during this time was to relate our background reading to the development of a theoretical framework taking account of our emphasis on language and meaning in the interpretation of our data. What we found was a picture of great richness but also of considerable complexity. In this paper we have chosen one particular episode that we found to be particularly rich. It was also very challenging to us to make sense of in terms of the starting point of our study, i.e. how a focus on language and meaning can assist students in reconstructing their algebraic knowledge.

The episode is presented with a view to discussing the evolution in our perspectives arising from the process of analysing the data collaboratively as a group and also with other colleagues. The reader is invited to make their own initial interpretations prior to this discussion.

The episode

This particular episode took place at the end of the first session with the 2-Year BEd students. They had been working on the Algebraic Pairs activities for the first part of the session and then had spent the latter part in a discussion of mathematical terms such as expression, equation, function, variable etc. As the session was almost complete, the researcher provided the opportunity for any questions, reactions or general discussion. The result was an extensive and articulate series of responses from one student in particular - Anthony (AG). Anthony is a mature student who had previously worked in industry as an engineer. Steve (ST) who is another mature student also makes contributions. This section of the transcript is essentially the result of a group interview situation in which Anthony is stimulated to reflect on his recent involvement in the group activity and also on his past mathematical experience.

1 **BH** OK, I was going to think about further activity but seeing as we've only five minutes
2 left, I think we'll end. Unless, are there any particular things that struck you as we've been
3 talking, that you want to return to, words which conjure up...
4 **AG** It's obvious as we start talking about maths, we start talking about functions, some
5 people have got a clearer view; that my image I realise now, when I'm teaching, I tend to
6 opt for, I like to see it as, that y = some function, it could be a = 3b plus something. I
7 keep returning to y = some function of x and if I saw it in a textbook for example that
8 $2(x+3)$ my automatic reaction would be to write $y = (2x+3)$ before I give it to the children
9 to do. $y = 2(x+3)$.
10 **BH** What would you be thinking of asking them to do next?
11 **AG** I'd be asking them to multiply the brackets out to give me a $y = 2x+6$ or asking them
12 to substitute a value of x and tell me what y is because that on its own as a function –
13 $x+3$. I suppose to me it is just floating about in mid-air with no relationship to anything.
14 It's totally intangible, what is it? what's it for? So if I ask them to multiply that bracket out
15 I got $2x+3$ before, now I've got $2x+6$, still doesn't lead to anything, doesn't mean anything,
16 doesn't tell me what it's from or where it's from, so my automatic reaction is to put the y =
17 in. Otherwise you've got that floating about and that is a function, then you've got
18 function. To me, what is a function,where does it come from, where does it come from?
19 **BH** You'd be happy to relate it to y. What would that mean then for you?
20 **AG** There's a missing number y and a missing number x. If we put any value in for y, or
21 any value for x... If we can find a value for y then we can find a value for x and if you get
22 into a quadratic there'd be two answers for y, so actually you're using something to solve a
23 problem.
24 **BH** Just taking that, say it was $y = 2x$ squared plus 6 times something...You said two
25 values.
26 **AG** Again, as soon as you get an x squared, I tend to think that it's probably going to be
27 two answers. Depends on...
28 **ST** There's only two answers to $y = 0$. You're only finding...What happened... that's your
29 equation of your line and you've got your $2x$ or your quadratic and everyone says there's
30 two answers or there could be two answers for a specific value of y - there's an infinite
31 number of answers, but there are only two answers for when $y = 1$. This is when $y = 1$.
32 **BH** Which is?
33 **ST** When $y = 0$.

34 ST That still gives a quadratic form. Yes. Or, you could say what are the answers, this is
35 when $y = 1$, if you wanted to and still use a...
36 AG When you've got $y =$, a quadratic function of x.
37 BH Yes.
38 AG When you've got that, there's a possibility of finding not only one answer for y but
39 the possibility of finding two answers for y.
40 BH Do you mean whenever y takes any value?
41 AG I'm coming from a realistic point of view in that I've got a specific problem of trying
42 to find out what this value of y is and in doing that I've made an equation in order to
43 solve my problem and in trying to solve my problem I might find that there are two values
44 of the x.
45 BH Say we had that? What about if I said y was minus 10?
46 AG Minus 10? Then there might not be a solution to it, no real solution. No solution to
47 my real world. This idea of no real solutions, you've gone into a hypothetical
48 world. You've gone out of a real life situation. From experience, in my situation, you've
49 gone out of a real life situation, you're going back to a hypothetical situation. You're going
50 right full back in circles to functions, that's something hypothetical, that's floating about,
51 not related to anything or solving anything. It's not come from any real life situation, it's
52 just a function, it's not related to anything else. I think that's why I have difficulty in
53 seeing where it's coming from.

Where we were

The literature that is referred to here is a sub-section of that which helped to formulate our thinking during the course of this study. The reason for choosing to highlight it at this point is that it was found to be the most pertinent to the initial analysis of the particular episode that is the focus of this paper. In fact we found that much of the research on algebra at the school level is relevant to this study e.g. that of Lesley Booth (1984), which includes studies concerned with school children's errors in algebra and an analysis of these errors leading to a categorisation of types of meaning associated with algebraic notation. Also of relevance is the work of Anna Sfard and Liora Linchevski (1994) who develop their *theory of reification* according to which there is an inherent process-object duality in the majority of mathematical concepts. They argue that the operational (process orientated) conception emerges first and that the mathematical objects (structural conceptions) develop afterwards through reification of the processes. They argue further that this is a difficult stage for many learners to achieve and that it is seldom accomplished quickly or without difficulty. They propose that the development of algebraic thinking is accomplished by means of a sequence of ever more advanced transitions from the operational to the structural. In particular they consider two especially crucial transitions: that from the purely operational algebra to the structural algebra 'of a fixed value' i.e. an unknown and then from there to the functional algebra of a variable. Further relevant research is that of Carolyn Kieran (1989), who emphasises the recognition and use of structure as a major area of difficulty in algebra. She distinguishes between *surface* structure and *systemic* structure. The former, in relation to an algebraic or arithmetic expression refers to the given form or arrangement of the terms and operations. The latter refers to the properties of the operations, such as commutativity and associativity, and to the relationships between the operations, such as distributivity (systemic is used in the sense of relating to the mathematical system from which the expression inherits its properties). She highlights the equality relationship between left- and right-hand expressions of equations as a 'cornerstone' of much work in teaching algebra and also that for some students, teaching methods based on the structure of equations do not succeed. She reports that for such students, who tend to view the right-hand side as the answer, 'the equation is simply not seen as a balance between right and left sides nor as a structure that is operated on symmetrically'. She also comments on how some students fail to see the surface structure of algebraic expressions and that this difficulty 'seems to continue throughout the algebra career of many students, as evidenced by errors such as reducing $(a+b+c)/(a+b)$ to c, seen amongst many college students'. In a further paper Carolyn Kieran and Nicholas Herscovics (1994) propose the existence of a *cognitive gap* between arithmetic and algebra that can be characterised as 'the student's inability to operate spontaneously with or on the unknown'.

An initial analysis of the episode suggested a number of links with the background literature and theoretical framework previously outlined. In order to help the reader make better sense of the transcript, it is worth emphasising at the outset that Anthony does not distinguish between the terms function and expression. In fact he refers to $2(x+3)$ as a function rather than as an expression at line 8. In relation to the theoretical framework and activity theory in particular, there are a number of references to a lack of purpose when dealing with functions. For example at line 14, Anthony asks 'what is it? what's it for?' and at line 15 says that it 'still doesn't lead to anything' and goes further to say that it 'doesn't <u>mean</u> anything'. This statement fits with Erik Schultz's interpretation of meaning as the 'purpose or intention of the cultural product' which in this case is the word 'function'. Anthony's description also suggests that he is working operationally for much of the time, e.g. at line 8, he says that 'my automatic reaction would be to write $y = (2x+3)$' and also at lines 16/17 he says that 'my automatic reaction would be to put the y = in'. His comments also suggest a lack of appreciation of the structural properties of equations e.g. at lines 11/12 he would 'be asking them (the children) to... substitute a value of x and tell me what y is' This suggests a view, consistent with the work of Carolyn Kieran, of 'the right hand side as the answer'. His comments at line 20 'There's a missing number y and a missing number x' suggest that he has not made the transition, in Anna Sfard's terms, from the 'structural' algebra of 'a fixed value' to the 'functional' value of a 'variable'. It seems from Anthony's comments that he sees the purpose of an equation as being to find a missing number and not to express a relationship. In Carolyn Kieran's terms, the equality relationship is not fully recognised i.e. the equation as a balance between right- and left-hand sides and as a structure to be operated on symmetrically.

To an extent these observations are typical of many students who have sought the help of the Academic Maths Support, although they were surprising to the researchers, as Anthony was seen to be a mathematically capable, though not strong, student. However much of what Anthony had to say was left untouched by this analysis and we were left with a sense of the inadequacy of the various theoretical frames, through which we had viewed our data, to account fully for what Anthony had to say. It seemed that there was evidence of resistance to 're-learn' algebra on Anthony's part, and that much that was being said related to his sense of identity and to his view of the nature of mathematics. None of this seemed to have been addressed in our first readings of the data. As a result of wider discussions with colleagues we decided to look to social practice theory for a 'wide(r) angle lens' (Dengate and Lerman, 1995) through which to view our data. In particular we turned to the work of Jean Lave and Etienne Wenger (1991) and that of Jean Lave (1996).

Lave and Wenger stress the essentially social character of learning and propose learning to be an aspect of a process of participation in socially situated communities of practice. They discuss the notion of Legitimate Peripheral Participation (LPP) which describes the particular mode of engagement of a learner in a new community of practice, whose level of participation is at first legitimately peripheral in the practice of the expert. The move from peripheral participation to full participation is seen as a dynamic process, characterised by changing levels of participation. Writing in 1996, Jean Lave describes the direction of movement as a *telos* and gives the example of 'becoming a respected, practising participant among other tailors or lawyers, becoming so imbued with the practice that masters become part of the everyday life of the Alley or the mosque for other participants and others in their turn become part of their practice'. She proposes that this might form the basis of 'a reasonable definition of what it means to construct identities in practice'.

Returning to the analysis of the transcript, it seems that there is considerable resistance on Anthony's part to re-construct his view of algebra. His view of a function is that 'it is totally intangible' (line 14) and 'with no relation to anything' (line 13). It is 'floating about in mid-air' (line 13), without meaning, e.g. 'what is it?' or purpose 'what's it for?' (line 14). It seems that Anthony's view of mathematics is only meaningful if 'you're using something to solve a problem' (lines 22/23). Having a problem to solve is real e.g. 'I'm coming from a realistic point of view' (line 41) and equations are simply tools to solve 'my problem' (line 43) e.g. 'I've made an equation to solve my problem' (lines 42/43). In formulating his views on the nature of mathematics, Anthony also seems to be saying significant things about his own sense of identity. His background is that of an engineer working in industry over many years and his path into Higher Education and teacher training would have been via vocational routes. Anthony seems to be calling on his previous experience (as expert) in this particular community of practice and also on his developing expertise in the practice of 'school teacher' to emphasise his identity as a part of the 'real world' e.g. 'my experience, my situation' (line 48). This contrasts with his view of the community of practice of mathematicians, as exemplified by the researcher, who inhabits 'a hypothetical world' (lines 47/48) and who has departed from the real world e.g. 'you've gone out of a real life situation' (line 48). He stresses his view that the researcher/mathematician is going nowhere, e.g. 'You're going right full back in circles to functions, that's something hypothetical - that's floating about, not related to anything or solving anything. It's not come from any real life situation, it's just a function, it's not related to anything else.' (lines 49-52). However he does seem to express some sympathy and desire for a greater level of participation in the practice of being a mathematician when he says 'I think that's why I have difficulty in seeing where it's coming from.' (lines 53/53). This also seems to reflect his peripheral participation in this particular community of practice.

It seems that our interest in language and meaning at the outset of this study has given us a picture of some of the ways in which our students are working on re-learning algebra. However, it also revealed much more - a complex set of phenomena and questions with which to revisit both our data analysis and also the ongoing development of our own practice. The process through which the evolution in our perspectives has taken place can be summarised as taking place in three broad stages. Firstly, we brought our initial focus on language and meaning to bear on our analysis of the data. Secondly we considered the broad background literature on the teaching and learning of algebra and took what we found to be resonant with our own study from that source. Finally in view of the fact that we felt that we had not fully accounted for the richness of the data with which we were presented, we turned to the 'wider angle lens' of social practice theory. The process has enriched our interpretation of the particular episode but more importantly has enriched our perspectives in general and will enhance our future practice as researchers in this area.

Acknowledgement

We would like to thank colleagues who attended seminars of the Mathematics Education Research Group and the Learning and Teaching Research Institute at Sheffield Hallam University during 1996 and also participants at the session at the BSRLM Conference in November 1996 for their contributions to the development of our thinking about this study. In particular we would like to thank our colleagues Professor Peter Ashworth and Mick Nott for their insights and contributions.

References

Booth, L.: 1984, *Algebra: children's strategies and errors*, NFER-Nelson.

Crawford, K.: 1996, 'Vygotskian approaches in human development in the information era', *Educational Studies in Mathematics* **31**, 43-62.

Dengate, R. and Lerman, S.: 1995, 'Learning theory in mathematics education', *Mathematics Education Research Journal* **7**(1), 26-36.

Elliott, S and Johnson, S.: 1995, 'Talking about algebra', *Proceedings of the Second UK Conference of Adults Learning Maths - ALM -2*, Goldsmith's College, London.

Kieran, C.: 1989, 'The early learning of algebra: a structural perspective', *Research Issues in the Learning and Teaching of Algebra*, Lawrence Erlbaum Associates, pp. 33-56.

Kieran, C. and Herscovics, N.: 1994, 'A cognitive gap between arithmetic and algebra', *Educational Studies in Mathematics* **27**, 59-78.

Kozulin, A.: 1990, *Vygotsky's Psychology: A biography of ideas*, Harvester.

Lave, J. and Wenger, E.: 1991, *Situated Learning: Legitimate Peripheral Participation,* Cambridge University Press.

Lave, J.: 1996, 'Teaching, as learning, in practice', *Mind, Culture and Activity* **3**(3), 149-164.

Lerman, S.: 1996, 'Intersubjectivity in mathematics learning: a challenge to the radical constructivist paradigm', *Journal for Research in Mathematics Education* **27**(2), 133-150.

Pimm, D.: 1995, *Symbols and Meanings in Mathematics*, Routledge, London.

Schultz, E.: 1994, 'The hermeneutical aspects of activity theory', *Activity Theory*, **15/16**, 13-16.

Sfard, A. and Linchevski, L.: 1994, 'The gains and pitfalls of reification - the case of algebra', *Educational Studies in Mathematics* **26**, 191-228.

Vygotsky, L. S.: 1962, *Thought and Language*, MIT Press.

MATHEMATICS, LANGUAGE AND DERRIDA

Tony Brown

Manchester Metropolitan University

Abstract

Derrida's revolutionary work in the study of language has seriously challenged the way in which we see words being attached to meanings. This paper makes tentative steps towards examining how his work might assist us in understanding the way in which our attempts to describe or capture our mathematical experiences modify the experience itself. In doing this we draw on the work of Jacques Derrida and John Mason in locating possible frameworks through which to conceptualise the relationship between language and mathematical cognition. It concludes that mathematical meaning never stabilises since it is caught between the individual's ongoing experience and society's ongoing renewal of its conventions. That is, mathematics, language and the human performing them are always evolving in relation to each other.

Introduction

Recently, there was a conference dedicated to the work of Jacques Derrida. Derrida was in attendance. A colleague and conference attendee, Antony Easthope, described how Derrida patiently sat through numerous papers speaking of his work without passing any comment. However, at the end of the conference Derrida made his own presentation. Having declared his delight to be at a conference celebrating his work he was, nevertheless, uncomfortable listening to so many people describing his work. He spoke of how, in attending the conference, he had experienced a sensation of being already dead. Having witnessed numerous attempts to sum up his work and integrate it elsewhere had made him feel as though his work had already been frozen for eternity, as if people were no longer seeking his present thinking.

In mathematics, as we survey the offerings of our students, maybe we are susceptible to the same sort of premature encapsulation, wrapping things up as we get some semblance of the thing we ourselves seek in their work, torn as we are between encouraging the gradual development of their own individual mathematical understanding, whilst ensuring they meet the social requirements of knowing certain specific ideas. Such difficulties can also arise as we attempt to capture the flow of our own mathematical thinking in to some sort of fixed form for the purposes of sharing. We can never fully express what we see. How indeed can mathematical thinking, a highly temporal commodity, be organised with reference to more stable entities, mental or otherwise? Can we freeze ideas, hold them still while we look at them? Addressing these issues is far from easy since the whole notion of finding stability is complex. In attempting to frame mathematical ideas what sort of loss do we experience?

This paper considers the relationship between mathematics and language, making particular reference to the work on language by Derrida. In taking this perspective I examine how language functions in organising mental activity and suggest that since language is so fundamental to the social formation and individual construction of mathematical ideas, it conditions all mathematical experience (cf. Brown, 1997). In this spirit I argue that linguistic reduction is an inevitable aspect of any mathematical construction, both locating and conditioning broader cogitations. Here the linguistic framing itself is intrinsic to the process of stressing and ignoring that underpins any conceptualisation. I consider this perspective in relation to that offered in Mason's work in the field of mathematics education and propose a partial reconciliation between these two perspectives.

I commence with a brief outline of the theoretical background I am assuming generally, and in mathematics education in particular, in addressing these concerns. I then introduce an outline of Mason's conception of language in developing mathematical understanding. This is followed by a more detailed account of Derrida's work before I identify some similarities in these two approaches that suggest an interesting way forward.

Theoretical background

Many contemporary writers on language, like Derrida, see the generation of language as instrumental in the self-formation of society and of the individuals within it. That is, language gives rise to the things we see as well as itself being a consequence of what we see. In this perspective, language can no longer be a seen as providing an unproblematic labelling of the world. Analytic philosophy's notion of language 'picturing' reality (e.g. Russell, 1914; early Wittgenstein, e.g. 1961) no longer holds up as an adequate metaphor for the way in which language functions, although such a belief may well still govern the everyday actions of many people.

How then can we see contemporary work in language as assisting us in discussing mathematical ideas? Like language, mathematics can be regarded as a socially produced phenomenon and, in particular, mathematical activity can be seen as being a form of linguistic performance. For Derrida social phenomena can always be read as a text, that is be understood as displaying a certain differential structure. Consequently, in this perspective, mathematical activity finds itself subject to the scrutiny of modern day critiques of language which emphasise its situation in history, in culture and in personal accounts. The human subject engaged in mathematics is positioned in a number of co-existing social agendas which flavour the style of engagement. Insofar as we see mathematical meaning being generated in the mind we cannot escape the formative influences on the mind. Also, we cannot partition off a section of the mind and label it 'mathematics'.

The dominant traditions of mathematics teaching have focused on how mathematics is, rather than on how it is seen. Teaching media are customarily treated as if they give access to something actually there. The parameters of mathematical activity are clearly delineated, where the symbols assume an unproblematic relation with the concepts they represent. Recent work in mathematics education research, however, such as that resulting from the widespread interest in constructivism, has focused more on how participants experience the mathematics classroom. Such 'insider' views of mathematics consciously build in some sort of self-reflective dimension. Since participants are necessarily governed by certain social practices I suggest such views are always embedded in a culture. That is, mathematics only manifests itself in activity governed by culturally specific norms (e.g. National Curriculum mathematics or university mathematics). It seems that insider perspectives are becoming more prominent as the absoluteness of mathematics itself is increasingly brought into question. Mason (e.g. 1994a, b) has spoken of researching problems, both mathematical and professional, from the *inside*. Constructivists have focused more on the individual learner's understanding of the mathematical tasks they face (e.g. von Glasersfeld, 1991) and gone on to consider this individual as a representative of the society from which he or she comes (e.g Ernest, 1997). Meanwhile, Skovsmose (1994) has commenced the groundwork in formulating a philosophy of critical mathematics education to examine ways in which discourses operate within mathematics education. Part of his task has been to uncover the way in which mathematics education conceals its intentions beneath the language it employs in declaring its project, such as in the way it gets presented as a subject where the learner's task is to be initiated into existing practices. He attempts to challenge this in coining the word *mathemacy*, "as a radical construct ... rooted in the spirit of critique and the project of possibility that enables people to participate in the understanding and transformation of their society" (Skovsmose, 1994, p. 27). It is in this spirit that I proceed.

Mathematics and language: a contemporary view

I shall begin by briefly looking at the position taken by Mason in associating mathematics with language, as an example of a recent writer in mathematics education moving away from assumptions of language picturing reality. It appears that he sees a clear distinction between mathematical experience and the linguistic description of it. The following quotes give a flavour of his view:

> Words generate more words in explanation, but often draw us away from the experiences from which they stem. (Mason, 1994a, p.176)

> Express to yourself in action (by doing it) and in words (by talking to yourself or a colleague) a rule for continuing the following array... (Honsberger quoted by Mason, 1989, p.3)

On the one hand we have the experience, on the other, the description of it in words. In my conversations with him, John Mason defends the content of his mind as not being reducible to description in words. Whilst I may report on my experience as a mathematician, in so doing I insert a gap between experience and report, resulting in the precise nature of my experience being rather elusive, being partly lost, at least as regards its capturing in language.

How then can we locate mathematical meanings in relation to mathematical and linguistic performance by humans? Traditionally, the task of the teacher and learner may be seen as sharing a pre-existing mathematics not susceptible to individual interpretation. Such an account is governed by notions where the teacher seeks to direct the student's attention to a specific way of seeing an objectively understood mathematics. Truth is embedded within the mathematics and the student seeks to locate this. Meanwhile, 'insider' views of mathematical problems, as exemplified in the work of Mason, focus on directing the student on a journey around problems and reflecting on the experience of doing mathematics. Whilst the emphasis is more on the student reconciling his or her experience with ways of describing it, Mason argues, the attempt to describe in words might draw the student away from the mathematical experience itself. The teacher's intention is rather less didactic but this may not necessarily imply a less conventional view of the underlying mathematics. Mason follows Gattegno in seeing truth gravitating around personal awarenesses, that is, truth is located in the mind of the individual.

So where is mathematics located? In more traditional views the mathematical meaning is independent of individual human performance. According to Mason however, the emphasis is on the individual human's personal awarenesses of mathematics. Nevertheless, both appear to see the description of mathematical activity in words as being outside the realm of mathematics itself. It is this that I wish to question.

Derrida's account of language

Contemporary accounts provide a view of language that infiltrates, whilst coalescing, the reality it serves. I wish to argue that the framing of mathematical experience in words by individuals should be seen as an integral part of the mathematics itself, inseparable from less visible cognitive activity. At this point I feel I need to go a little deeper into the theoretical apparatus that I will utilise in presenting my argument. I shall briefly mention the hermeneutics of Ricoeur and Gadamer as a prelude to developing Derrida's more radical post-structuralist position.

Ricoeur (e.g. 1981) and Gadamer (1962) assert that experience itself is conditioned by the language we introduce to describe it. For them, language mediates truth. That is, whilst they are happy with the notion of truth, they see this truth as being obscured by our attempts to access it and describe it. The chief consequence of this for our current analysis is that mathematical experience and description of it in words are drawn closer together. Indeed, for Ricoeur the meaning of an event is closely related to how it is described (Thompson, 1981, p. 126). The material existence of the world is fully accepted but it only presents itself according to some particular way of organising what is seen. The material world lights up as it is touched by the human's gaze. As an example, I was rather surprised when a cousin of my wife evaluated the picturesque Peak District landscape we were walking through from a military perspective. He was seeing completely different things to those that I would see in my normal guise as an unathletic rambler motivated by nice views and proximity to a pub. A "pretty red bush" to me was a "point of cover overlooking the valley entrance" for him.

Objectivity itself is historically created, defined in terms of the way in which the individual consciousness perceives the material. This partitioning of the material world into phenomena is closely related to the descriptions made in respect of it. Mathematical objects then, present within such thinking, are not unproblematic entities for all to see, but rather, are understood differently by each individual. The distinction between such phenomena and the perception of them is softened with phenomena and perception evolving together through time. In this perspective mathematical ideas, as located through notation, are not endowed with a universal meaning, but rather, derive their meaning through the way in which an individual attends to them. For example, if someone is shown the expression:

$x^2 + y^2 = 1$

it may be seen as mixture of numbers and letters with no particular significance, as an algebraic equation, as a representation of a circle, or *as* a circle. Thus mathematical 'object' and human 'subject' are seen in a more complementary relation as part of each other. The emphasis in this formulation is on the individual's experience of grappling with social notation within his or her physical and social situation. This provides a framework, seen from the individual's point of view, in which the distinction between the individual and the social is softened. In building his or her understanding, the individual is obliged to work through the social filter of language. My strategies for making sense of and acting in the world are always underpinned by cultural stylising derived through language, whether I be mountaineering with a climbing guide, dancing in the latest fashion or doing mathematics. All such activities can be seen as specific "discursive spaces" (Stronach and Maclure, 1996, p. 262). Derrida (1978, pp. 278-293) similarly claims that you can only observe others using language from the home base of your own particular way of using language. One is always positioned within culturally derived ways of seeing and so experience itself is understood through inherited structuring embedded in language usage. For example, I am currently participating in a research project which focuses on initial training students on the mathematical component of a primary education course. Confronted with numerous and often conflicting discourses on the art of teaching mathematics, the students weave a version for themselves about what mathematics is, what its components are, how it might be taught etc., which bears little resemblance to the ways in which members of the research team, or indeed the Chief Inspector of Schools, would mark out the 'same' territory. They use words in different ways, choose to highlight different things and generally understand their task in what seems to be a completely different way. Yet as a research team member I can only use my own way of seeing and describing their situation as my starting point, which perhaps I can change with a little effort, but *my* understanding of their perspective is always conditioned from where *I* start out.

Language is always used to describe a world already conditioned by language. Any human performance can be read as a 'text' in the philosophical sense of the word (as a differential structure). Indeed, Derrida famously asserts (1976, p. 158), "there is nothing outside of the text", nor are there truths to provide points of anchorage. He sees differential structures as being inherent in explicit language, consciousness and unconsciousness. Like Lacan (1977), he identifies this as a general feature of the mental world, with both conscious and unconscious being structured like a language; a sequence of differentiated terms. The mental world, so seen, is a system of differences, part of which is claimed by explicit linguistic structuring (Derrida, 1982). It should be stressed, however, Derrida does not dismiss the experience itself; rather, experiences are in a constant state of flux conditioned by attempts to associate them with a never-ending linguistic flow. He would see mathematical involvement as necessarily textual, brought about through human partitionings of the world - a framing that is, in a sense, already there, consequential to the cultural linguistic heritage (see for example Derrida, 1989).

> My own words take me by surprise and teach me what I think. (Merleau-Ponty, quoted by Derrida, 1978, p. 11).

Derrida builds on this quote in discussing how inscription in words (and maybe also in symbols, in diagrams) orients psychologically produced phenomena. If I may risk using his own, rather slippery, words:

> If writing is inaugural it is not so much because it creates, but because of a certain absolute freedom of speech, because of the freedom to bring about the already there as a sign of the freedom to augur. A freedom of response which acknowledges as its horizon the world as history and the speech which can only say: Being has already begun... (Writing) creates meaning by enregistering it, by entrusting it to an engraving,

> a groove, a relief, to a surface whose essential characteristic is to be infinitely transmissible. Not that this characteristic is always desired, nor has it been; and writing as the origin of pure historicity, pure traditionality, is only the telos for a history of writing whose philosophy is always to come. (Derrida, 1978, p. 12) [1]

I take Derrida to mean, very crudely, that inscription in writing functions closely in relation to the psychological phenomena it locates and, indeed, becomes part of it. In reading Derrida myself, I never get to what he means but rather I experience the ongoing sensation of being moved on before I am ready. His words never frame the final version of his 'present' thinking. In this respect, Derrida's position is not that far away from the the more moderate line of Gadamer and Ricoeur who permit an ongoing renewal within the co-evolution of phenomena and perception. However, Derrida's refusal to allow any anchorage in truth makes his work quite distinctive and all the more radical in its ability to reject orientation around universal structures. Taking such a view within mathematics, we would thus be more aware of being oriented around historically-generated ideas flavoured by their situation within human enterprises. Derrida addresses this explicitly in his first book which examines Husserl's work *The Origin of Geometry* (Derrida, 1989).

Derrida's position takes language well beyond its traditional scope towards embracing the whole of human experience. Objections sometimes arise when we attempt to nudge language into this extended domain. For example, on the surface at least, such views appear unsatisfactory to those who wish to defend the power of their own mathematical experiences as being outside the realm of language. Nevertheless, whilst Mason might pursue such a line he does speak of manifestations in the 'outer' which have some sort of association with 'inner' experience. He suggests that we need to acknowledge "a world of experience that is not material, not phenomenal, but inner, with access through what we are able to read in the outer" (Mason, 1994b, p. 7). This, I feel, invites a degree of compatibility between his understanding and the line taken by Derrida.

We need, however, to ask about the nature of this association between inner and outer and question how these outer manifestations attach themselves to inner experience. Are they like the tips of icebergs (i.e. part of the thing being signified) or like road signs (i.e. separate to the thing being signified)? I suggest words, diagrams and other manifestations of mathematical activity, can be seen as functioning in either way, according to current interest and the emphasis one assumes. Indeed they may be seen as two points on the hermeneutic cycle connecting ways of experiencing and ways of describing. The physical environment, for example, is textual, in Derrida's sense, insofar as the human eye organises it differentially. Thus 'seeing' is always in relation to prior conditioning. Any attempt at inscription reflects this broader but maybe more

---

[1] N.B. *augur* - to foretell through signs, to guess or conjecture; *inaugural* - to cause to begin; *telos* - aim, purpose, ultimate end.

elusive differentiality. As with Saussure (1966), Derrida sees the signifier/signified duality as inseparable. But as with Lacan (and Wilden, 1968), Derrida sees relatively stable signifiers being associated with a fluid underbelly, comprising a signified field which sweeps out to occupy the whole of consciousness, and indeed, the unconscious. Both presence and absence are located by the signifier. Any loss incurred in the attempt to articulate remains attached to the signifier seeking to replace it. The signifiers are haunted by what they squeeze out. Meanings are derived only through retrospective examination of the flow of signs. The component signifiers do not have implicit meanings, only relational associations with other signifiers in the chain. There are no independently existing meanings in the chain since any attempt to frame in words, any attempt to 'mean', creates a gap between 'being' (through time) and attempts to explain it (in a fixed form). Lacan speaks of an indefinite sliding of meaning to convey the "impossibilities" of attaching one word with one meaning. We have no truths to provide orientation apart from those generated through this system of differences that is forever changing. Derrida (1981, translator's introduction, p. ix) suggests that self-present meanings are illusions brought about through repressing the differential structures from which they spring. However, as a note of caution, Derrida seems to have back-tracked a little from the extreme way of thinking many associate with him:

> ... it was never our wish to extend the re-assuring notion of text to a whole extra-textual realm and to transform the world into a library by doing away with all boundaries, all frameworks, all sharp edges. (Derrida, 1991, p. 257).

Mason meets Derrida?

In describing mathematical experience, what are we describing? Presumably some sort of cognitive activity which structures the grey matter somehow. It is this sort of structuring which Derrida sees as differentiality rather like that which we find in a chain of words. With language seen in this way, it becomes harder to draw a clear distinction between the experience and the description of it. We may suspend the 'presence' of the experience, in a sense, but the experience itself was understood differentially and thus already a suspension, so no more nor less the 'real' experience. There is no experience outside the text (or differential structure), only a retroactive construction of it asserted by the individual. To make strict distinction between experience and description of it in words, as Mason appears to, requires a relatively restrictive view of language where words act only like road signs. Whilst Mason's distinction might offer a valuable rhetorical device in initiating or analysing mathematical performance, it may squeeze out the picture the linguistic forces driving (and being driven by) the mathematical construction (cf. Derrida, 1989).

An alternative account of language sees the spoken word as rather more like the tip of an iceberg, that is, as part of the thing it signifies, in particular, the silent cognitive activity taking place around it. Both hermeneutics and post-structuralism, offer accounts which, through being more flexible in their understandings of language, engage with the material qualities of the world. They accept the material world but intercept perception before it assumes shared notions of categorising this material world into objects. To varying extents for these writers, language itself is the home base and so subjects meet in their shared use of the manifestation of this in speech or writing and associations between language and reality resist stability between signifier and signified. Rather, both reality and language are caught in an historical process of mutual formation which is never complete, nor even pauses long enough for one to map the other.

Within such a formulation, I suggest the historicity present in both the genesis and the current performance of mathematics is recognised. For example, our perception of school geometry has changed through the popularisation of computer packages such as *Cabri*. Our orientation has changed as a consequence of being able to work differently on particular features that resisted analysis in a pencil and paper environment. Both the key features of geometry and how we are working with them are evolving. Thus, the learning of mathematics moves away from being concerned with recreating existing ideas but instead emphasises the tightly-knit relation between language and understanding and how they move on together. Mathematical constructing, I would suggest, is always linguistic to a degree, oscillating in a hermeneutic circle, between more or less sturdy linguistic packaging. In this sense there is no mathematics outside language.

So then, whilst Mason seems to be sharing with more conventional work in mathematics education a preference for seeing language as a labelling device he is less willing to suppose that teachers can share their own mathematical images with their students. In assuming the teacher's task himself, Mason is concerned with enabling his students to generate their own mathematical experiences. That is, he does not explain the mathematics in his head but rather, initiates an activity which he hopes will enable his students to experience some mathematics and in this, perhaps, encounter certain ideas. He sees learning as a journey of self discovery:

> ...it is important to re-search, re-collect, re-connect, re-learn, re-integrate, and re-cast insights in the discourse of the times. I see working on education not in terms of an edifice of knowledge, adding new theorems to old, but rather as a journey of discovery and development in which what others have learned has to be re-learned, re-integrated and re-expressed in each generation. (Mason, 1994a, p. 177).

On this point Mason and Derrida seem close. For Mason, ideas are not inherited prepackaged and intact, but rather, each new generation will engage in tasks that give rise to new understandings of what might be seen as old ideas. There is a need to work on ideas, they cannot just be 'received'. This way of thinking bears a striking similarity with some of Derrida's recent work:

> Inheritance is never a *given*, it is always a task ... there is no backward looking fervour in this reminder, no traditionalist flavour, Reaction, reactionary or reactive are but interpretations of the structure of inheritance. That we *are* heirs does not mean that we *have* or that we *receive* this or that, some inheritance that enriches us one day with this or that, but that the *being* of what we are *is* first of all inheritance, whether we like it or know it or not. (Derrida, 1994, p. 54).

So Mason shares with older traditions the idea of language being separate to that which it describes. However, he sides with Derrida in supposing that we can only ever have new understandings of old ideas.

Conclusion

In this paper I have attempted to demonstrate how contemporary writing about language offers valuable insights that may assist us in understanding how language intervenes in developing mathematical understanding. I have suggested that the framing of mathematical experience in words by individuals should be seen as an integral part of the mathematics itself. As such language does not merely label mathematical experience but rather activates it and shapes it. And in this process both language and mathematical experience get re-evaluated. That is, the mathematical features, around which we orient our work, evolve as well as our way of working with them. Whilst the student's task may well oscillate between fitting language to mathematical experience and bringing meaning to language through reflection on one's own experience, both experience and linguistic production forever continue, resisting attempts to settle on a particular version. Mathematical meaning never stabilises since it is caught between the individual's ongoing experience and society's ongoing revision of conventional mathematical practices. Mathematics, language and the human performing them are always evolving in relation to each other. There is no final version to be learnt, since we lack universal truths to hold this in place. Recent research in mathematics education, such as that referred to here, has begun to prepare us for the consequences of this recognition.

I conclude with a quote from another post-structuralist thinker, writing shortly after the student uprising in Paris in 1968:

> Just as psychoanalysis, with the work of Lacan, is in the process of extending the Freudian topic into a topology of the subject (the unconscious is never there in *its* place), so likewise we need to substitute for the magisterial space of the past- which was fundamentally a religious space (the word delivered by the master from the pulpit above with the audience below, the flock, the sheep, the herd) - a less upright, less Euclidean space where no one, neither teacher nor students, would ever be in *his final place* (Barthes, 1977, p. 205).
>
> In short, within the very limits of the teaching space as given, the need is to work patiently tracing out a pure form, that of a *floating;* a floating which would not destroy anything but would be content simply to disorientate the Law. The necessities of promotion, professional obligations .. , imperatives of knowledge, prestige of method, ideological criticism - everything is there, but floating. (*ibid.* p. 215).

Acknowledgement

I am grateful to John Mason, Peter Gates, Ian Stronach and, in particular, Dave Wilson for extensive comments on earlier drafts.

References

Barthes, R.: 1977, *Image, Music Text,* Fontana/Collins, Glasgow.

Brown, T.: 1997, *Mathematics Education and Language: Interpreting Hermeneutics and Post-Structuralism,* Kluwer Academic Publishers, Dordrecht. The Netherlands.

Derrida, J.: 1976, *Of Grammatology,* Johns Hopkins, London.

Derrida, J.: 1978, *Writing and Difference*, University of Chicago Press, Chicago.

Derrida, J.: 1981, *Dissemination,* University of Chicago, Chicago.

Derrida, J.: 1982, *Margins of Philosophy,* Harvester Wheatsheaf, London.

Derrida, J.: 1989, *Introduction to Husserl's "The Origin of Geometry",* University of Nebraska Press, Lincoln.

Derrida, J.: 1991, 'Living on the border lines', in Kamuf, P. (ed.) *A Derrida Reader: Between the Blinds,* Columbia University Press, New York.

Derrida, J.: 1994, *Spectres of Marx,* Routledge, London.

Ernest, P.: 1997, *Social Constructivism as a Philosophy of Mathematics Education*, SUNY, Albany , NY, USA.

Foucault, M.: 1972, *The Archaeology of Knowledge*, Routledge, London.

Gadamer, H.G.: 1962, *Truth and Method,* Sheed and Ward, London.

von Glasersfeld, E. (ed.) : 1991, *Radical Constructivism in Mathematics Education,* Kluwer Academic Publishers, Dordrecht. The Netherlands.

Habermas, J.: 1972, *Knowledge and Human Interests*, Heinemann, London.

Lacan, J.: 1977, *Ecrits: A Selection*, Tavistock/Routledge, London.

Lacan, J. and A. Wilden: 1968, *Speech and Language in Psychoanalysis*, Johns Hopkins, Baltimore.

Mason, J.: 1989, 'Mathematical abstraction as the result of a delicate shift of attention', *For the Learning of Mathematics* **9**, 2-8.

Mason, J.: 1994a, 'Researching From the Inside in Mathematics Education: Locating an I-You Relationship', *Proceedings of PME 18*, University of Lisbon **1**, 176-194.

Mason, J.: 1994b, 'Researching From the Inside in Mathematics Education: Locating an I-You Relationship', An expanded version of a plenary presentation to *PME 18*, University of Lisbon, Centre for Mathematics Education, Open University, UK.

Ricoeur, P.: 1981, *Hermeneutics and the Human Sciences*, Cambridge University Press.

Russell, B.: 1914, *Our Knowledge of the External World,* George Allen and Unwin, London.

Saussure, F. de.: 1966, *A Course in General Linguistics*, McGraw-Hill, New York.

Schutz, A.: 1962, *The Problem of Social Reality,* Martinus Nijhoff, The Hague.

Skovsmose, O.: 1994, *Towards a Philosophy of Critical Mathematics Education,* Kluwer, Dordrecht.

Stronach, I. and Maclure, M.: 1996, 'Mobilising meaning, demobilizing critique? Dilemmas in the deconstruction of educational discourse', *Cultural Studies* **1**, 259-276.

Wittgenstein, L.: 1961, *Tractatus Logico-Philosphicus*, Routledge, London.

This paper is adapted from Tony Brown's book *Mathematics Education and Language: Interpreting Hermeneutics and Post-Structuralism,* published in the Mathematics Education Library series of Kluwer Academic Publishers in 1997. It is printed here with permission.

STUDENTS' INTERPRETATIONS AND REPRESENTATIONS OF NUMBER

Declan O'Reilly

University of Sheffield

Abstract

The findings of major research studies in this country and the United States leave little doubt as to the difficulties which many students experience in dealing with number – difficulties which persist well into the secondary school (Hart et al., 1981; Foxman et al, 1985; Carpenter et al, 1981). A review of the literature also reveals an accumulating body of knowledge concerned with the relationship between the processes children use and the concepts they acquire (Hiebert and Lefevre, 1986; Hiebert and Wearne,1986, 1992; Nesher, 1986). In particular, research has used terms such as 'reification' or 'encapsulation' to describe the 'process-object' relationship (Sfard, 1991; Gray and Tall, 1994). The work described here, which was conducted entirely off-computer, was part of a larger project, one aim of which was to investigate how the 'process-object' relationship would be mediated through working in the computational medium of Boxer.

Background

The study reported here formed an intermediate part of a larger study (O'Reilly, 1995). In the first phase of that research, students learned how to program in Boxer (see the Appendix). In the latter phase, they utilised their programming knowledge to learn about number. This study acted as link between these phases insofar as it established the students' interpretations and representations of number.

The Research Study

The larger study described above took place in an Inner London primary school over a period of two years (O'Reilly, *op. cit.*). Four pairs of case study students, chosen to represent ability and gender, were tracked through years 5 and 6, i.e. between the ages of 9 and 11. At the outset, the children had no experience of Boxer, and little or no experience of Logo in particular, or computing in general. Throughout this time, I spent an average of one day a week in the school as a participant observer. Within the classroom, they had access to a single computer which was connected to a laser printer. For the on-computer sessions, data were recorded in the form of systematic video recordings along with annotated screen print-outs. For the off-computer sessions, copies of the children's written work were made and the follow-up discussions were audio recorded.

The Interviews

This phase of the research consisted of a series of semi-structured interviews with two groups of year 6 students: the case study students (later joined by two other Boxer students) and a comparison group of eight students from a parallel class. Students were first interviewed in pairs, where they were asked to interpret and represent numbers. Then, in a subsequent interview (in groups of about four), they were invited to comment on the representations. These interviews had two main aims: to establish baseline data on the students' understanding of number, and to gain insight into the representations that they employed.

The following account is in two parts. The first part is concerned with the students' interpretations of quantities obtained through extending the number system. The second part focuses on their representations of these quantities.

Students' Interpretations of Number

Natural Numbers

By way of a gentle introduction to the interviews, students were simply asked to choose any three numbers, and were then asked to select two of these numbers (Q1a and Q1b).

Q1a: Choose any three numbers and write them down.

Q1b: Now choose two of these numbers and write them down.

All of the students chose whole numbers, which would seem to suggest that, for them, the concept of number was equated with whole number. Moreover, most of the numbers were relatively small (all but one were within the range 1 - 100), suggesting further that the numbers selected were those within their experience. There were no discernible differences between the Boxer group and the comparison group of students (see Table 1).

Non-Boxer		Boxer	
Annabel	**10, 11,** 15	Neil	**3, 9,** 8
Tony	10, **15, 40**	Julia	3, **8, 15**
Brian	**3,** 5, **9**	James	**1,** 6, **11**
Helen	**3,** 20, **69**	Leroy	6, **19, 210**
Alan	**9, 15,** 20	Christine	**5,** 24, **18**
Clare	3, **6, 9**	Lesley	**1, 2,** 3
Rose	**10,** 120, **3**	Jacky	**2, 4,** 9
Derek	1, **100, 5**	Kirsty	**9, 12,** 55
		Rita	**12, 6,** 22
		Susan	7, **3, 2**

The final two numbers chosen are in bold.

Table 1: Students' choice of whole numbers

Rational Numbers

Question 1, besides being intended to help the students feel at ease, was also meant as a lead-in to Question 3, which probed their awareness of non-integer quantities. As part of the protocol, I continued to ask this question until the students gave two successive integers. They were then asked to choose a number between these. In contrast to the responses to Question 1, there were very marked differences between the Boxer and non-Boxer groups' answers to this question.

Q3 a, b, c, etc: Can you write down a number that is bigger than ... [the smaller number] and smaller than ... [the bigger number]?

A frequent first response at this point was for the non-Boxer students to deny that there were any numbers between two whole numbers. Their later fractional answers, which followed researcher interventions, reinforced the contention that the term 'number' was seen – at least initially – as synonymous with whole number. It was very noticeable that each of these students gave fractional answers and that each of these fractions was obtained by employing a bisection strategy. For example, Derek's first response to this question was to offer 5½ as a number between his two integers 5 and 6. His next, and final response, was 5¼. It suggested that their notion of fractional quantities was tied to the physical operation of halving.

The initial responses of the Boxer group indicated an awareness of a number continuum. Five of the ten Boxer students gave a decimal expression as a number in between two whole numbers. This was very different from the non-Boxer students, none of whom gave a decimal answer to this question. Julia, one

Non-Boxer			Boxer		
	Consecutive whole numbers	First & Last non-integers		Consecutive whole numbers	First & Last non- integers
Annabel	**10, 11**	10½, 10¼	Neil	**3, 4**	3.5, 3.487
Tony	**15, 16**	15½ , 15¼	Julia	**8, 9**	8½, 8.05
Brian	**3, 4**	3½ , $3\frac{1}{8}$	James	**1, 2**	1½, $1\frac{1}{5}$
Helen	**3, 4**	3½, 3¼	Leroy	**19, 20**	19.5, 19.01
Alan	**9, 10**	9½, nil	Christine	**5, 6**	5.5, 5.1
Clare	**6, 7**	6¾, 6½	Lesley	**1, 2**	1½, $1\frac{1}{8}$
Rose	**3, 4**	4½, 4¼	Jacky	**2, 3**	2½, 2¼
Derek	**5, 6**	5½, 5¼	Kirsty	**9, 10**	9½, nil
			Rita	**6, 7**	6.5, 6.05
			Susan	**2, 3**	2½, 2¼

Table 2: Students' choice of non-integers

of the Boxer students, whose first answer was 8½, converted her second fraction 8¼ to 8.25. Her subsequent responses of 8.23, 8.10 and 8.05 as numbers between 8 and 8.25 indicated a quite sophisticated conception of the real number continuum.

In a later question (Q.9), I again probed the students' appreciation of decimals, and again there were marked differences between the two groups.

Q9a: Do you know about decimals?

Q9b: Which is the bigger 0.5 or 0.13?

All of the comparison group claimed to know about decimals, but only two out of the eight chose 0.5 as bigger than 0.13 (Table 3). At the time of the interviews, the non-Boxer students had begun to learn about decimals in the context of money, and this may account for the disparity between their two answers.

Eight of the ten Boxer students chose the correct answer 0.5. But, again some caution needs to be exercised in interpreting their answers. Most of the them recognised 0.5 as a half, but several of them had difficulties in explaining why 0.13 was a smaller quantity. This may be because this is a difficult question to answer orally. I did not simply accept the Boxer students' answers as construing self-evident knowledge of decimals, but rather continued to probe the rationale for their choice.

Kirsty, who chose 0.13, said: "But, I think I'm wrong." Christine, who chose the correct option, asked if it was a 'trick question', and could not explain why 0.5 was bigger than 0.13. Jacky chose 0.5 and knew it was a half, but she could not explain why it was bigger than 0.13. Lesley justified her choice of 0.5 by saying:

Non-Boxer			Boxer		
	9a	9b		9a	9b
Annabel	yes	0.13	Neil	yes	0.5
Tony	yes	0.13	Julia	yes	0.5
Brian	yes	0.13	James	yes	0.13
Helen	yes	0.5	Leroy	yes	0.5
Alan	yes	0.13	Christine	yes	0.5
Clare	yes	0.13	Lesley	yes	0.5
Rose	yes	0.13	Jacky	not sure	0.5
Derek	yes	0.5	Kirsty	not sure	0.13
			Rita	yes	0.5
			Susan	yes	0.5

Table 3: Students' knowledge of decimals

"The lower it is, actually the higher it is." The reasoning for this classic 'LS' (larger is smaller) error (Foxman et al, 1985) could be inferred from what she said a little while later, i.e. "That's a half and that's a thirteenth."

Directed Numbers

Question 5, like Question 1 for decimals, was intended as a lead-in question for further investigation of students' conceptual understanding of directed numbers.

Q5: What is the smallest number that you know?

Often a first response by the non-Boxer students to this question was to ask "Is zero a number?". Assured that it was, many chose it as their smallest number. For some, there also seemed to exist a quantity '-0', which may have arisen because they had heard of temperatures below zero, although this is speculation on my behalf. Six of the eight chose 0 or -0, with Rose's 0 -¼ being a variation on this. A frequent first response by the Boxer students was to ask: if they could use 'minuses'. Five of the ten did so, choosing quantities ranging from Kirsty's -1 to Julia's minus sign followed by her depiction of a infinite number of nines. Of the remaining five students, three chose decimals and two chose 0.

The question arises as to why the responses of the two groups should have been so different? Neither of the groups had been exposed to negative numbers as part of the formal school curriculum. However, the Boxer students were coming into contact with both decimals and directed numbers in the course of their work in the computer environment. These findings suggested that these contacts were indeed affecting their expression of number. They appeared to have both a wider and a deeper insight of numbers than their non-Boxer counterparts. Two follow up questions (Q6 and Q8) reinforced this impression.

Non-Boxer		Boxer	
Annabel	-0	Neil	1000000000000000000 000
Tony	-0	Julia	-99999999999999999 999999999999 . . .
Brian	0	James	0.1
Helen	-infinity	Leroy	0.00001
Alan	0	Christine	0
Clare	0	Lesley	-infinity
Rose	0 -¼	Jacky	0
Derek	-10 000 000 000	Kirsty	-1
		Rita	-50
		Susan	0.1

Table 4: Students' choice of smallest number

Q6a: Can you think of a cold temperature?

Q6b: Can you give me a colder temperature?

The responses to Question 6 appeared to contradict the answers to Question 5, in that many of the students who gave 0 as the smallest number they knew were now writing of numbers below zero. However, three points may be proffered as partial explanations. Firstly, there is the obvious one of temperature being a meaningful context: it is not claimed that they knew which number was smaller but rather which temperature was colder. Secondly, it could be that the term 'smallest' is ambiguous when negative numbers are involved, i.e., students may have been giving the smallest absolute value $|x|$ of a number x. Thirdly, 0 may be the smallest number that students have *experienced*.

Several of the students in both groups used non-standard notation to depict the relative values of negative numbers in the context of temperature.

Non-Boxer			Boxer		
	6a	**6b**		**6a**	**6b**
Annabel	20 below 0	30 below 0	Neil	0^{c}	-10^{c}
Tony	1 c	1/2 c	Julia	0^{c}	-1^{c}
Brian	3 c	-1c 0	James	-40^{0} C	-60^{0} C
Helen	-0 c	-20c	Leroy	1^{0} F	0.1^{0} F
Alan	-0	Freezing	Christine	1^{0} C	0^{0} C
Clare	-1	-0	Lesley	4^{c} +	-22^{c}
Rose	0 - 1/8 c	0 1/24 c	Jacky	1 or 2 F^{0}	0 F^{0}
Derek	0 c -	100c -	Kirsty	5FH	below 1
			Rita	-10	-30
			Susan	0.1	below freezing

Table 5: Students' comparison of directed numbers

Another Question (Q8) was designed to elicit further information concerning students' conceptions of both decimals and directed numbers in terms of the real number continuum.

> Q8: Suppose the temperature changed from -1 to 1. What numbers would it pass through?

The qualitative differences in the responses to this question by the non-Boxer and Boxer groups were quite stark (see Table 6 opposite). Brian's and Helen's answers though non-standard (-00 00, and -0 respectively) could be interpreted to mean 0 and therefore be correct, and likewise, Rose's response '0 1/2'. Of the remaining five, three gave a sequence of numbers above 1, and two gave a single number below -1.

Two of the ten Boxer students, Christine and Kirsty, gave incorrect responses. Of the remaining eight students, seven gave two or more fractional or decimal quantities. James, for instance, wrote down all the numbers between -1 and 1 obtained by using 0.1 increments. On the other hand, it seems clear that there was some confusion with the relative sizes of negative numbers, particularly if these were not whole numbers. Julia, who gave only positive decimal numbers in her answer, was asked whether she could write down any numbers between -1 and 0. The difficulty she had in doing so can be gauged from her comments: "I don't know. If you write a number like one point something that's making it smaller, because you go down in the negatives. But if you make it zero point something, that's above zero."

Non-Boxer		Boxer	
Annabel	2 6 8 10	Julia	0.05 0.10 0.15 0.25 0.50 0.75
Tony	5 7 3 8	Neil	-.6 -.5 0 .3 .4 .5
Brian	-00 00	James	-1 -0.1 -0.2 -0.3 -0.4 -0.5 -0.6 -0.7 -0.8 -0.9 0 0.1 0.2 0.3 0.5 0.6 0.7 0.8 0.9 1
Helen	-0	Leroy	00.1 00.2
Alan	-2	Christine	-2 -4 -6 -8 -10 +
Clare	-7	Lesley	-¼ $-{}^{1}/_{8}$ -½ -1 +¼ +½ $+{}^{1}/_{8}$ $+{}^{1}/_{16}$
Rose	0 ½	Jacky	½
Derek	+1 +50 +90 +Hot	Kirsty	1½
		Rita	0.5 0.05
		Susan	(a) +½ +¼ +¾; (b) -½ -¼ -¾

Table 6: Students' interpretation of the real number continuum

Students' Representations of Numbers

Reacting to a conference on representation, Belanger (1987) warned of the dangers of neglecting students:

> I have felt during our deliberations that students are strangely absent from our discussion; this is similar to the period of the 1960s when they were absent from the new math and the new science and there is a danger they will be missing from the new representations. One of the things we need to remember is that students construct representations. (p. 105)

In the light of Belanger's warning, this section can be seen as constituting my efforts to ensure that students' contributions were not ignored.

Natural Numbers

Having chosen two whole numbers for Question 1, the students were asked if they could illustrate their choice with drawings.

> Q2: Suppose you were trying to explain to a younger child which of your two numbers was the bigger, can you make a drawing that would help them to understand?

Their responses to this question indicated that the relationship between what was represented (the signified) and the mode of representation (the signifier) was more complex than had previously been supposed. It seemed that for some students, it was sufficient to depict the 'bigness' of one number relative to the other - they felt no need to accurately preserve the size of each number. Secondly, there was some ambiguity as to whether the representations were intended to accurately represent each of the two numbers or simply depict the approximate relationship between the two. In the follow-up discussions, the students themselves critiqued such limitations in their own and others' representations.

Helen, who had drawn the picture of jumpers (Figure 1a), reacted to Michael's preference for it by pointing out what she thought was its weakness.

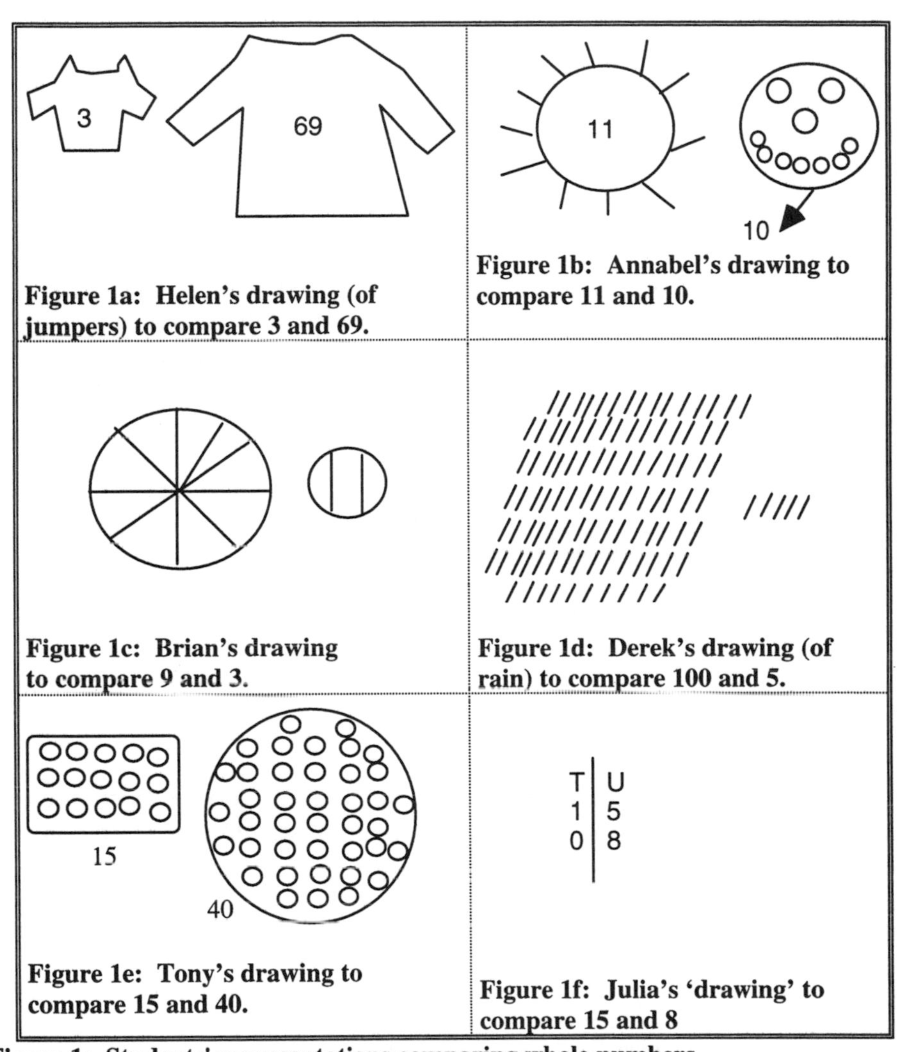

Figure 1a: Helen's drawing (of jumpers) to compare 3 and 69.

Figure 1b: Annabel's drawing to compare 11 and 10.

Figure 1c: Brian's drawing to compare 9 and 3.

Figure 1d: Derek's drawing (of rain) to compare 100 and 5.

Figure 1e: Tony's drawing to compare 15 and 40.

Figure 1f: Julia's 'drawing' to compare 15 and 8

Figure 1: Students' representations comparing whole numbers

Speaker	Dialogue	Commentary
Michael:	I like this one. I like that because it shows size.	Points to the drawing of jumpers. [Figure 1a].
Helen:	I don't think that's a very good idea.	
Interviewer:	You don't think your own idea is very good now. Why not?	
Helen:	Because if it was a bigger number, this would be like gigantic.	Refers to the difficulties in keeping the drawings in proportion to the numbers.

Extract 1

Figure 1a was also criticised by another child because "It is not clear what the numbers stand for." Other students criticised Figure 1b because "One is for summer and one is for winter, but what's that got to do with 10 and 11?" Figure 1c was thought defective because "The parts aren't the same", but there was disagreement between Rose and Clare about the merits of Figure 1d (the raindrops).

Speaker	Dialogue	Commentary
Interviewer:	[to Rose] Why do you not like it so much?	
Rose:	It would be hard to count.	
Clare:	Yes, but even if it is hard to count, they'll still see the difference because of the amount of raindrops.	Refers to depiction of relative sizes.

Extract 2

Leroy, who thought the drawing of stones (Figure 1e) was good, was countered by Julia, leading to a discussion on whether a diagram should be able to convey its meaning without the need for back-up writing.

Speaker	Dialogue	Commentary
Leroy:	That one. Fifteen is smaller than forty.	Points to Figure 1e.
Julia:	Yea, but just looking at that, you couldn't tell the difference between thirty-nine and forty.	
Leroy:	But you can write the numbers next to them.	
Julia:	But, in that case, you can just write them.	

Extract 3

Julia's implicit use of an abacus mirrors a common teaching approach to place value. Before drawing/writing it, she asked "Do you want a 'drawing-drawing', or can I write it?" This hints at an intermediate interpretation of this representation, i.e., between that of a drawing and the purely symbolic form of a number. Jacky, chose this diagram saying: "If they understand about tens and units, that one would be quite good". Similarly, Rita responded, "If children already knew a lot about numbers, that one would be a good idea."

Rational Numbers

Just as Question 2 was intended to evoke representations to illustrate the two whole numbers from Question 1, Question 4 was meant to fulfil the same role with respect to the rational numbers from Question 3.

> Q4: Suppose you were trying to explain to a younger child which of your two numbers was the bigger, can you make a drawing that would help them to understand?

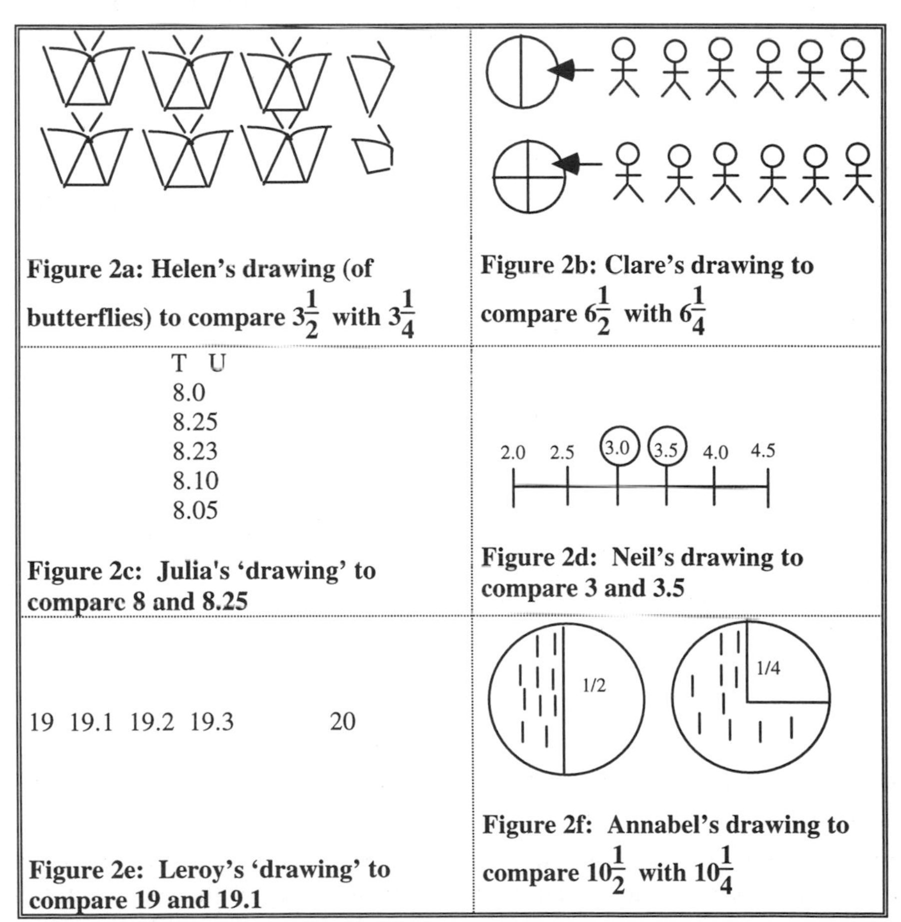

Figure 2a: Helen's drawing (of butterflies) to compare $3\frac{1}{2}$ with $3\frac{1}{4}$

Figure 2b: Clare's drawing to compare $6\frac{1}{2}$ with $6\frac{1}{4}$

Figure 2c: Julia's 'drawing' to compare 8 and 8.25

Figure 2d: Neil's drawing to compare 3 and 3.5

Figure 2e: Leroy's 'drawing' to compare 19 and 19.1

Figure 2f: Annabel's drawing to compare $10\frac{1}{2}$ with $10\frac{1}{4}$

Figure 2: Students' representations comparing rational numbers

Often students chose representations for fractions which were similar to those used to compare whole numbers, even when these are clearly inappropriate. An answer which included half a person or half a car seemed to be regarded as quite acceptable. Helen was the exception to this. She later described her own diagram (Figure 2a) showing portions of butterflies as "Disgusting". Students also mixed representations within the same drawing, as in Clare's case (Figure 2b), where stick people were used for whole numbers but fractions were represented by parts of a circle. Another phenomenon observed was for students to represent the fractional parts only, ignoring the whole number part. In general, the students experienced some difficulties in conjuring up a representation for non-whole numbers (Derek, for example, could not come up with one). This may be due to the novelty of the idea, i.e. representing numbers rather than manipulating them.

The phenomena of using inappropriate representations, of mixing representations, and of depicting only the fractional parts were also observed amongst the Boxer students. But these students tended towards representations which depicted all numbers and a general trend towards more symbolic notation. Julia and Neil consistently chose conventional classroom representations: Julia the implicit abacus model (Figure 2c) and Neil the number line (Figure 2d). Although he did not draw a number line, Leroy's depiction of numbers in a line (Figure 2e) was clearly equivalent, as can be gauged from the following extract.

Speaker	Dialogue	Commentary
Leroy:	You have nineteen and twenty there.	Writes 19 on paper, leaves a large space and then writes 20.
Leroy:	Now, in between these two numbers, there are lots of littler numbers. No, not littler. There are lots of other numbers. First, there is nineteen point one.	Writes this.
Leroy:	. . . and then it goes up like this to ten.	Writes 19.2 and then 19.3.
Leroy:	But then there are more littler numbers. No, more numbers. They, go up like this.	Counts up 19.4 19.5 etc.
Leroy:	They go on to ten which is twenty.	Refers to next step in the sequence beyond 19.9.

Extract 4

This clearly means that 19.9 + 0.1 = 20, but note how counting orally with decimals might proceed: nineteen point eight (19.8), nineteen point nine (19.9), and then nineteen point ten (19.10).

In the follow-up interviews with both Boxer and non-Boxer students, criticisms of other representations tended to be linked to the accuracy of the representation of the fractional part. Rosie's comment on Figure 2a was typical: "Because if you were trying to show a small fraction, like a tenth, you wouldn't be able to see it."

Directed Numbers

As explained earlier, questions about directed numbers were asked in the context of temperature, and so the request for representations were similarly framed (Q7):

Q7: Suppose you were trying to explain to a younger child about temperature, can you make a drawing that would help her or him to understand ?

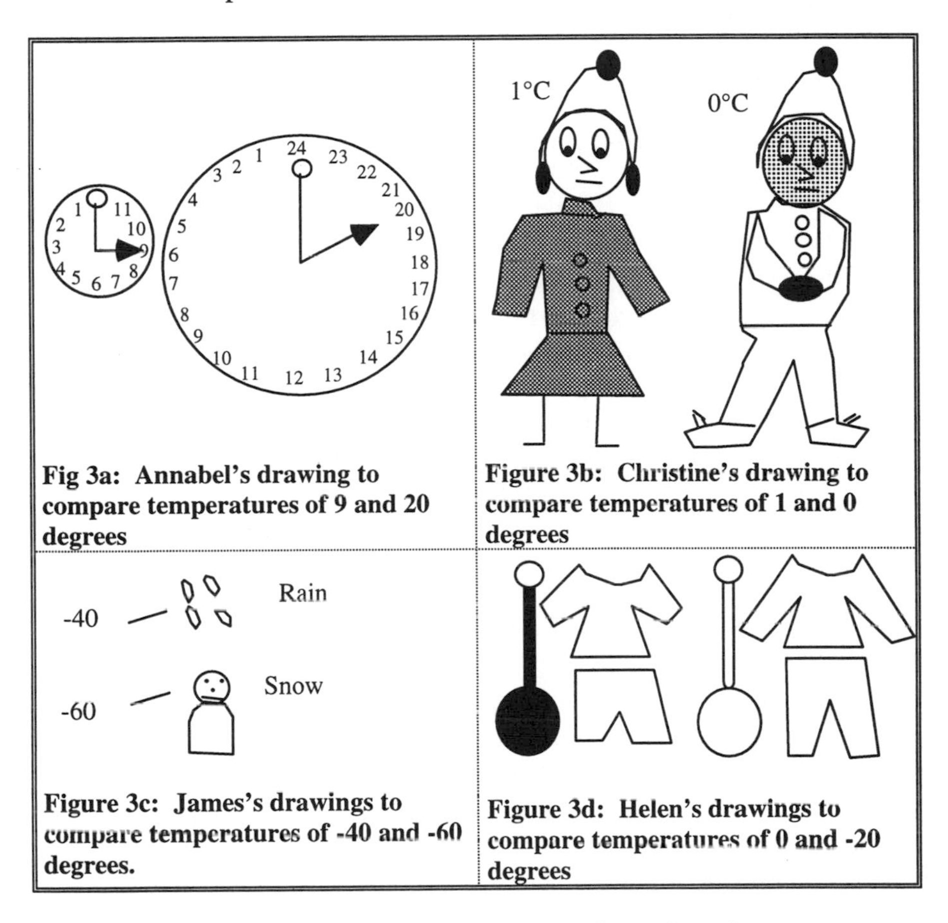

Fig 3a: Annabel's drawing to compare temperatures of 9 and 20 degrees

Figure 3b: Christine's drawing to compare temperatures of 1 and 0 degrees

Figure 3c: James's drawings to compare temperatures of -40 and -60 degrees.

Figure 3d: Helen's drawings to compare temperatures of 0 and -20 degrees

Figure 3: Students' representations comparing directed numbers

Not surprisingly, many students drew thermometers, but often these were graduated in a way that gave no clear indication of the difference between their two numbers, or not graduated at all. In this sense, they might best be considered as labels for temperature. Helen's drawings of summer and winter clothes (Figure 3d) alongside thermometers to denote the difference between her two temperatures of 0 and -20 degrees lends weight to this labelling conjecture. The use of light and warm clothing was also used by Christine (Figure 3b), who added 'cold' and 'warm' shading to her images. James's juxtaposing of numbers next to images of rain and a snowman (Figure 3c) again draws attention to one of the deficiencies of this form of representation, namely the lack of clarity concerning the nature of the relationship between the number and its accompanying drawing.

Summary and Implications

This paper has analysed the responses of two groups of 10- to 11-year-old students to a series of questions concerning their interpretations and representations of whole numbers, negative integers and non-integers. These two aspects of number are now discussed in terms of the findings and their implications for mathematics education in general.

Interpretations

The findings show that there were few problems with whole numbers. However, this was not the case for non-integers. Pimm (1987) observes that there is no such thing as a decimal number or a fractional number *per se*: i.e. 2.5 and 2½ are simply different representations of the same number. However, my questions designed to probe students' interpretations of such numbers suggested that the relationship between signifier and signified is not a simple one. Fractional notation, used predominantly by the comparison group, carried with it an action-based connotation of its derivation, i.e., 'a half of', 'a quarter of'. 'In-betweeness' could thus be signified by such expressions. The greater use of decimals by the Boxer group appeared to suggest a greater appreciation of a number continuum, but attaching meaning to decimal symbolism was problematic for both sets of students.

Investigating directed numbers in the context of temperature demonstrated the advantages of a meaningful context for all of the students. There were some conceptual problems in deciding whether zero was a number, and some unorthodox notation to depict temperatures below zero. The greater use of directed numbers by the Boxer group again pointed towards a greater appreciation of a number continuum, but there was some confusion with the relative sizes of negative numbers, particularly if these were not whole numbers.

The models used in elementary school textbooks to depict number vary according to both the type of number and the operation being modelled. Not unnaturally, there is an implicit assumption that if students are learning about number operations, then there already exists a base level of understanding about the numbers themselves. The evidence of this study challenges that assumption.

It shows that, while students did not naturally use either directed numbers or decimals, contexts could be found which elicited the use of the former. In the latter case however, similar contexts very often elicited the use of fractions rather than decimal expressions, particularly from the non-Boxer students.

Furthermore, the students' own explanations seemed to point towards understanding gained as a result of operations rather than preceding them. In the number phase, this led to the adoption of an 'operational approach' to number, i.e., the operations were the means by which the numbers themselves were first created. One of the interesting findings of that research was that children were inclined to describe numbers such as 0.2, not as 'two tenths' but as 'two divided by 10' – a description which retains its operational derivation. It is suggested here that approaching directed numbers and decimals through operations may help students to make connections between number processes and number concepts, i.e., develop their 'proceptual thinking' (Gray and Tall, 1994).

Representations

The findings show that students' representations of numbers tended to be linked to the particular numbers. In some cases, it was not clear whether diagrams were intended to stand alone, and in others it was uncertain what aspect of number was signified by the diagram. Where whole numbers and rational numbers were mixed, it was often the case that students also mixed representations: one aspect of the diagram to denote the whole number and another to denote the rational part. Representations for directed numbers were generally quite different from those for positive rational numbers. The desirability of having a single representation for all numbers was not generally appreciated.

Care must be taken that, in our critique of their constructions, we do not adopt a patronising stance. The limitations which students identified in their models of numbers alert us to deficiencies in our own conventional representations. It is common, for example, to use parts of rectangles or circles to depict fractions, but it is questionable whether such a representation could be easily extended to include large numbers, and it is not at all apparent how directed numbers might be represented in this form. Representations such as the number line may appear to overcome some of these difficulties, but such representations may not be as obvious as their familiarity suggests (Vergnaud and Errecalde, 1980; Vergnaud, 1983). The implications for teaching I would suggest are twofold: firstly, the need for teachers to recognise and make explicit the limitations of any representation, and secondly the benefits that may be gained by encouraging children to construct and discuss their own representations of number.

References

Belanger, M.: 1987, 'Excerpts from the conference', in Janvier, C. (ed.) *Problems of Representation in the Teaching and Learning of Mathematics,* Lawrence Erlbaum Associates, NJ, pp. 99 - 108.

Carpenter, T., Corbitt, M., Kepner, H., Lindquist, M. and Reys, R.: 1981, 'Decimals: Results and Implications from National Assessment, *Arithmetic Teacher* **28**(8), 34 – 37.

diSessa, A.: 1995, 'The many faces of a computational medium: teaching the mathematics of motion'. in diSessa, A., Hoyles, C., Noss, R. and Edwards, L. (Eds) *Computers for Exploratory Learning*. Springer-Verlag, Berlin, pp. 337 - 359.

diSessa, A.,Abelson, H. and Ploger, D.: 1991, 'An Overview of Boxer', *Journal of Mathematical Behaviour* **10**(2), 3 - 15

Foxman, D., Ruddock, G., Joffe, L., Mason, K., Mitchell, P. and Sexton, P.: 1985, *A Review of Monitoring in Mathematics 1978 - 1982*, HMSO, London.

Gray, E. and Tall, D.: 1994, 'Duality, ambiguity, and flexibility: a "proceptual" view of simple arithmetic', *Journal for Research in Mathematics Education* **24**(2), 116 – 140.

Hart, K. (ed.): 1981, *Children's Understanding of Mathematics 11–16*, Murray, London.

Hiebert, J. (ed.): 1986, *Conceptual and Procedural Knowledge: The case for Mathematics,* Lawrence Erlbaum, Hillsdale, NJ.

Hiebert, J. and Lefevre, P.: 1986, 'Conceptual and procedural knowledge in mathematics: an introductory analysis', in Hiebert, J. (ed.), *Conceptual and Procedural Knowledge: The case for Mathematics,* Lawrence Erlbaum, Hillsdale, NJ, 1 - 24.

Hiebert, J. and Wearne, D.: 1986, 'Procedures over concepts: the acquisition of decimal number knowledge', in Hiebert, J. (ed.), *Conceptual and Procedural Knowledge: The case for Mathematics,* Lawrence Erlbaum, Hillsdale, NJ, 199 – 223.

Hiebert, J. and Wearne, D.: 1992, 'Links between teaching and learning place value with understanding in first grade', *Journal for Research in Mathematics Education* **23**(2), 98 - 122.

Nesher, P.: 1986, 'Are mathematical understanding and algorithmic performance related?', *For the Learning of Mathematics* **6**(3), 2 - 9.

O'Reilly, D.: 1995, *Visualising Number: A Study of Children's Developing Sense of Number in the Computational Medium of Boxer,* Ph.D. Thesis, Institute of Education, University of London.

Pimm D.: 1987, *Speaking Mathematically,* Blackwell, Oxford.

Pimm D.: 1995, *Symbols and Meanings in Mathematics,* Routledge, London.

Sfard, A.: 1991, 'On the dual nature of mathematical conceptions: reflections on processes and objects as different sides of the same coin', *Educational Studies in Mathematics* **22** 1 - 36.

Vergnaud, G.: 1983, 'Why is an epistemological perspective a necessity for research in mathematics education', *Proceedings of the Fifth International Conference for the Psychology of Mathematics Education,* Montreal, pp. 2 - 20.

Vergnaud, G. and Errecalde, P. et al: 1980, 'Some steps in the understanding and the use of scales by 10 - 13 year-old students', *Proceedings of the Fourth International Conference for the Psychology of Mathematics Education,* Berkeley, pp. 285 - 291.

Appendix 1

Boxer

Boxer is an example of a new generation of software, being developed by Andrea diSessa and colleagues at the University of California in Berkeley (diSessa, 1995; diSessa et al, 1991). The name 'Boxer' stems from the fact that programs are typed inside boxes on the screen (Figure A1). There are two main types of boxes: 'doit' boxes (programs) and data boxes which carry information. Readers familiar with Logo will immediately recognise the genesis of Boxer in that language.

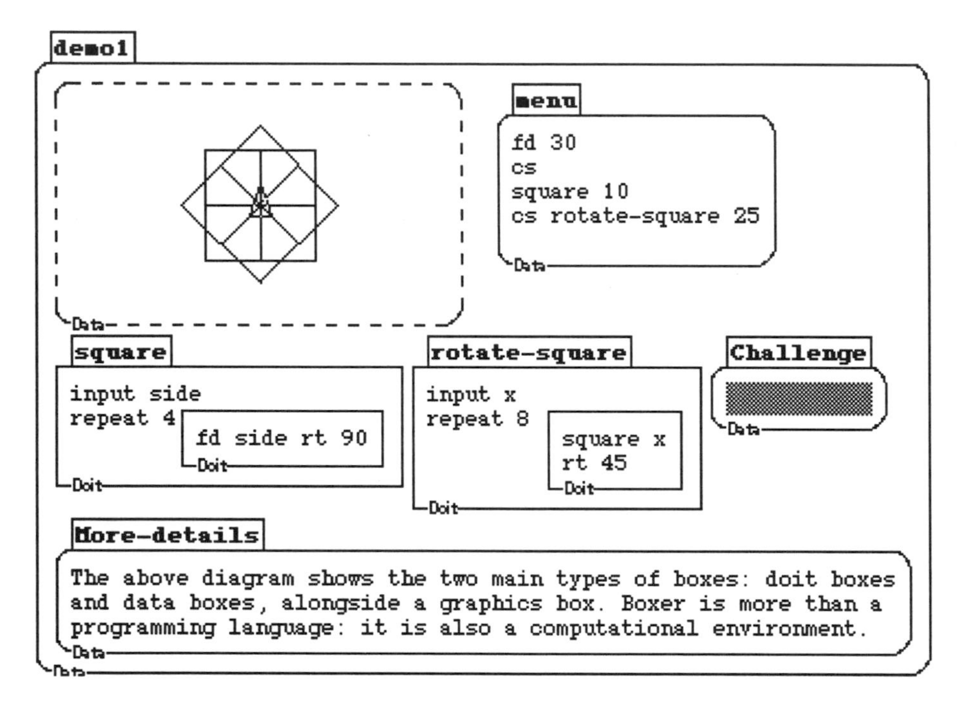

Figure A1: The boxes of Boxer

THE TRANSITION FROM GCSE TO A-LEVEL IN MATHEMATICS: A PRELIMINARY STUDY

Dylan Wiliam*, Margaret Brown*, Daphne Kerslake, Sue Martin, and Hugh Neill**

King's College, London* and University of Bath**

Abstract

An analysis of the grades obtained by the age-18 cohort who took A-level examinations in 1994, matched with their results obtained at GCSE two years earlier, supports the idea that the 'gap' between GCSE and A-level is larger for mathematics than for English, except at the highest levels of attainment. The difference in the size of the gap is approximately one-half of an A-level grade for candidates who attain grades B and C and approximately one grade for candidates who attain A-level grades D and E. Analysis of a small sample of GCSE and A-level scripts suggests that algebraic manipulation represents a particular difficulty for many students, although staff responsible for admission to A-level courses believe that these weaknesses can be overcome in the introductory parts of A-level programmes.

Introduction

This paper reports some of the results of a brief study that attempted to shed light on the size and nature of the 'gap' between GCSE and A-level in mathematics, in order to inform the review of qualifications taken by 16- to19-year-olds (Dearing, 1996a; 1996b; 1996c).

The notion of a 'gap' is, of course, deeply problematic. There is no obvious metric for measuring learning, and even if there were, the differences in curricula followed by students would render it meaningless, or at best, impossible to apply in practice. Nevertheless, there has been a widespread perception that many students find the transition from GCSE to A-level in mathematics more difficult than is the case in other subjects. Therefore, for the purposes of this study, the notion of a 'gap' was taken to involve the differences between the standards of achievement required at GCSE and at A-level, both in terms of the standards of the respective examinations, and in terms of the skills needed by students to engage with the respective courses.

The study was commissioned by the School Curriculum and Assessment Authority in September 1995, with a requirement to report within two months, and undertaken on behalf of the Joint Mathematical Council of the United

Kingdom. The restricted timescale necessitated a heavy reliance on the use of existing data, and the resources available limited the amount of original data that could be collected. For this reason, the study must be regarded as exploratory. However, we believe that the study's findings are suggestive, and may prompt others to investigate further.

The study incorporated three strands of evidence. The first strand (undertaken by Dylan Wiliam, with the assistance of Geoffrey Howson and Jason Tarsh) compared the *size* of the gap between GCSE and A-level in mathematics with that in English and other subjects, while the second (co-ordinated by Hugh Neill, and undertaken by Hugh Neill, Sue Martin, Dylan Wiliam, Roger Porkess, Mundher Adhami, Sam Boardman, Sue Chandler, Neville Hunt, Daphne Kerslake, Sue Pope and Teresa Smart) attempted to examine the *nature* of the gap. The third strand (co-ordinated by Daphne Kerslake and undertaken by Daphne Kerslake, Sam Boardman, Theresa Smart, Sue Burns, Dylan Wiliam and Roy Ashley) elicited responses from those involved in admitting students to A-level programmes, partly in order to validate the findings of the first two strands, but also to explore their *consequences*. The full report was published by SCAA in March 1996 (Wiliam, 1996).

Is the gap bigger in mathematics?

Table 1 shows the results of all candidates in the age-18 cohort who sat an examination in single A-level mathematics in 1994. Of 22,693 candidates who took A-level in 1994 after having obtained a grade A in GCSE mathematics in 1992 or earlier, just over a quarter (26%) achieved a grade A at A-level, just under a quarter (23%) got a grade B and 18% were awarded grade C. Grades D and E were obtained by 13% and 8% respectively of those with an A grade at GCSE. In total, therefore, 89% of those taking single A-level mathematics after getting a grade A at GCSE gained some kind of a pass in mathematics at A-level, although only two thirds (67%) got a 'good grade' (i.e. C or higher). (A similar database was available for A-levels taken in 1993. However, this included GCSE results only for those pupils who took GCSE in 1991, whereas many of the ablest students taking A-level in 1993 had taken GCSE in 1990 or even earlier. The result of this limitation was that the best predictor of A-level grade in 1993 appeared to be 'no result' at GCSE!)

Candidates doing single A-level mathematics after getting only a B in mathematics at GCSE fared much less well. Less than a third (32%) got a good grade and only just over two-thirds (68%) got any kind of pass at all. For those candidates attempting A-level mathematics after achieving only a grade C at GCSE, the prognosis was even worse: one-sixth (17%) achieved a good grade and only one-half (51%) got any grade at all. When one considers that many students with a C in mathematics are dissuaded, or prevented, from studying A-level mathematics at all (many schools and sixth-form colleges require that students have at least a grade B, or a grade C on the Higher tier at GCSE, before being allowed to study mathematics at A-level), then grade C in mathematics at GCSE is not a particularly good predictor of success at A-level mathematics.

This approach excludes students who entered for double mathematics A-level, but unfortunately the readily available data for double A-level mathematics gave only a single combined score (i.e. 10 for AA, 9 for AB, 8 for AC or BB, etc). One way of incorporating these data is to assume that the grades obtained by students entered for double mathematics differ at most by one grade and that it is the highest grade that is the most representative. The result of combining these data with those for single A-level is shown in Table 2.

		GCSE grades			
		A	B	C	D-G
	A	26	4	2	7
	B	23	11	5	13
A-level	C	18	17	10	19
Grades	D	13	19	15	18
	E	8	17	18	12
	X	11	32	49	31
	Total	100	100	100	100
	Numbers	22693	12009	3310	314

Table 1: Cross-classification (%) of grades for single A-level mathematics and GCSE mathematics

		GCSE Grades			
		A	B	C	D-G
	A	32	5	2	9
	B	23	11	5	13
A-level	C	17	17	10	19
Grades	D	12	19	15	18
	E	7	17	18	12
	X	10	31	49	29
	Total	100	100	100	100
	Numbers	26461	12224	3334	326

Table 2: Cross-classification of grades (%) for all A-level mathematics and GCSE mathematics

Even though the addition of the double-mathematics candidates changes very little, which of Tables 1 and 2 is the most representative is a difficult question.

Candidates doing double mathematics will typically have had twice as much teaching, which should increase their grade. On the other hand, the double mathematics students *are* part of the cohort, and, often being the most mathematically gifted students, would perhaps get an equally good grade had they studied only single mathematics at A-level. The balance of the argument would appear to favour the inclusion of the double mathematics candidates, and so for the purposes of the study, Table 2 was taken as the most representative for the transition from GCSE mathematics to A-level mathematics.

Comparisons with other subjects

The foregoing analysis suggests that candidates with GCSE mathematics grades B and C do not fare well at A-level. This could be taken as evidence of a 'gap', but the results are difficult to interpret in isolation. For this reason, similar analyses were conducted for the transition from GCSE to A-level in English, and also with results aggregated across all A-level subjects.

The transition from GCSE to A-level in English

The situation in English is complicated somewhat by the existence of two separate GCSEs - English Language and English Literature - and by the existence of two separate A-levels with the same titles. This yields a total of four possible routes from GCSE to A-level. However, the number of candidates taking English language at A-level is rather small and the results table for GCSE language to A-level literature is very similar to that for GCSE literature to A-level literature, and so the latter was taken as the basis for comparison with mathematics, and is shown as Table 3.

As in mathematics, a grade A in English at GCSE confers a very high probability of success in the subject at A-level, although in English, possession of a grade B at GCSE does appear to confer a greater advantage over a grade C than is the case in mathematics.

		GCSE grades			
		A	B	C	D-G
	A	32	6	1	1
	B	30	17	6	3
A-level	C	21	27	16	11
grades	D	11	27	29	23
	E	4	16	28	30
	X	2	8	21	32
	Total	100	100	100	100
	Numbers	19411	18081	8062	600

Table 3: Cross-classification of grades for A-level English literature and GCSE English literature

The transition from GCSE to A-level

The difficulty with a comparison of two subjects is that we don't know which is 'out of step'. Within the time-scale available there was not enough time to obtain and evaluate results for a sufficient range of other subjects to provide a 'base-line' for comparison. However, the *DFE's Statistical Bulletin 4/95* does cross-tabulate the average GCSE points scores obtained by students in the age-16 cohort in 1992 with the A-level points scores of the same candidates in 1994. Although in a far from ideal form, these data can be used (with suitable assumptions) to provide a 'base-line' showing how GCSE grades are converted into A-level grades. The detailed calculations are given in Wiliam (1996), but are summarised below.

The first stage was to equate GCSE points scores with average GCSE grades. In order to do this, some assumption about the average number of GCSEs taken by each student needs to be made. Given that this cohort was subject to the requirement to teach all the National Curriculum subjects for a 'reasonable time' in Key Stage 4, and given that this study excludes all students who did not sit at least one A-level within two years of taking GCSEs, it seems reasonable to assume that each student took an average of 10 GCSEs (see p. 44 for a discussion of the robustness of this assumption). Points scores were equated with grades by grouping all points scores to the nearest 'flat' profiles (e.g. 10 grade Bs).

The second stage was then to equate the A-level points score with A-level grades. This was complicated by the fact that not all candidates took the same number of A-levels. In general, higher-achieving students took more A-levels. For each GCSE grade, A-level points scores for 'typical' candidates at grades E, D, C, B and A were generated by multiplying the number of A-levels taken by candidates with that GCSE grade by 1, 2, 3, 4 and 5 respectively. Points-score boundaries were situated mid-way between the 'typical' scores.

The number of candidates in the points-score ranges were then allocated to grades in proportion to the amount of the points-range falling within that grade's range, resulting in Table 4. In order to make comparisons easier, the data from Tables 2, 3 and 4 are shown as a cumulative frequency graph in Figure 1.

		GCSE grades			
		A	B	C	D - G
	A	32	21	5	3
	B	32	26	13	5
A - level	C	22	23	18	13
grades	D	11	19	31	23
	E	2	7	21	31
	X	1	3	12	24
	Total	100	100	100	100
	Numbers	11818	50913	61714	33601

Table 4: Cross-classification of candidates' grades for all A-levels and all GCSEs

Interpreting the data in Table 4 is difficult, because of the number and the strength of the assumptions made, but the simplest interpretation is probably to conceive of it as showing how, on average, GCSE grades get converted into A-level grades. Of course, since student performance at GCSE is highly variable, and as students tend to select their better subjects for A-level study, Table 4 does not represent how GCSE grades would be converted into A-level grades for all students for all subjects. Furthermore, the data in the table is highly sensitive to the assumption about the number of GCSE subjects taken. For example, if the average number of GCSEs taken by A-level candidates had been taken as 9, rather than 10, then Table 4 would have been much more like the results for English Literature (Table 3). Clarification on this point is necessary before any weight is placed on the data in Table 4 and Figure 1. However, since students doing A-level mathematics or English are unlikely to have chosen these subjects unless they were among their best subjects, Table 4 probably does represent a reasonable basis for comparison with Tables 2 and 3.

Comparative analysis

Each of the cumulative frequency polygons (henceforth called 'curves' for convenience) in Figure 1 (opposite) is characterised by two features. One is the 'level' of the curve: the curve for the grades obtained by candidates with grade A GCSE English is above that for the grades obtained by candidates with grade B GCSE English (which is of course what we would expect). The other feature is the inclination of each of the curves - the three curves for mathematics seem to be somewhat 'flatter' than the other six curves.

The simplest approach is to characterise the 'level' of the curve by how far along the horizontal axis the curve crosses the median (i.e. the 50th percentile), and its inclination by the gradient at this point. If we regard the five A-level grades as being underlain by a continuous scale (with A=5, B=4 and so on), we can derive quantitative estimates of the 'level' of each curve, together with a measure of the spread of the A-level grades achieved (standard deviation). These data for the nine distributions are summarised in Table 5.

	Mathematics			English			Overall		
	A	B	C	A	B	C	A	B	C
level (grade at median)	4.2	2.1	1.1	4.4	3.0	2.1	4.4	3.9	2.5
slope (% points per grade)	22	19	18	29	27	29	32	23	31
mean A-level grade	3.31	1.74	1.12	3.68	2.47	1.62	3.79	3.26	2.14
sd of A-level grades	2.66	2.38	1.88	1.56	1.71	1.46	1.25	1.76	1.79

Table 5: Summary data for the cumulative frequency curves.

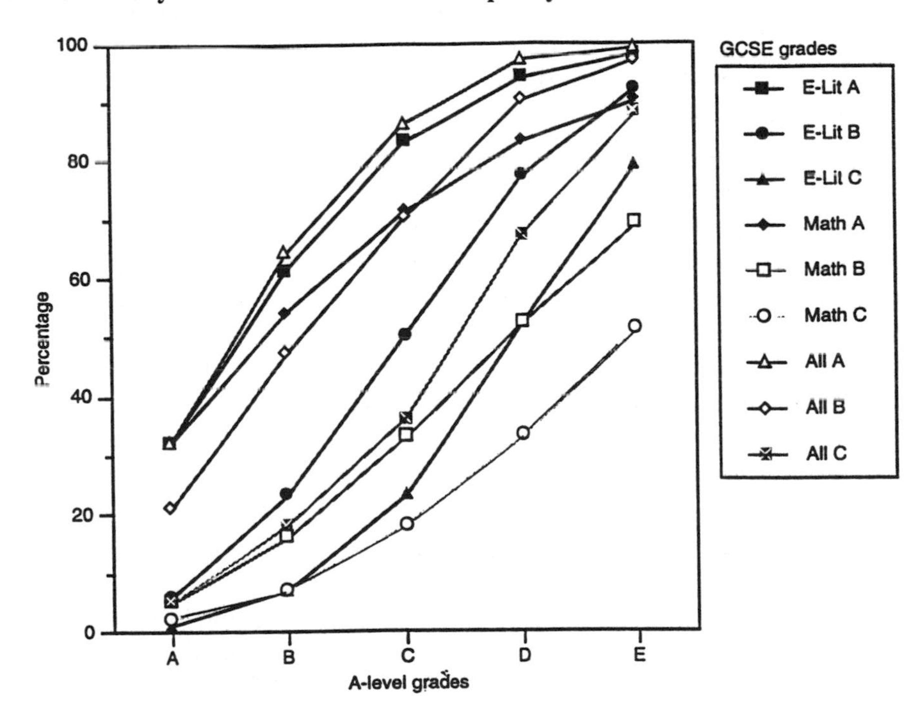

Figure 1: Cumulative frequencies of average A-level grades by average GCSE grades

Figure 1 and Table 5 between them summarise all the data generated for this strand of the study. There are many potentially interesting features, although it would be unwise to place too much reliance on these data, since they are based on only a single cohort and very strong assumptions were needed to generate the aggregate data for the overall GCSE–A-level transition. Nevertheless, several features are apparent from the data.

Approximately one third of A-level candidates who had a GCSE grade A get an A-level grade A, whatever the subject. For the highest-attaining candidates, therefore, it would appear that there is little evidence that the GCSE/A-level gap is any bigger in mathematics than in other subjects.

However, with the lower A-level grades, the picture is rather different. For example, to have at least a 50% chance of getting a grade D or better at A-level, one needs a grade B at GCSE in mathematics, but only a grade C in English. In a very real sense, therefore, the gap between GCSE and A-level is one grade bigger in mathematics than in English for candidates who attain A-level grades D and E. Interpolating between the graphs in Figure 1 suggests that for candidates who attain A-level grades B and C, the gap between GCSE and A-level is one-half of a grade bigger in mathematics than in English. Comparison between mathematics and *all* A-levels is difficult because of the sensitivity of the analysis to the assumption about the average number of GCSEs taken by the A-level candidates. Nevertheless, it does seem fair to conclude that the gap between GCSE and A-level is bigger in mathematics than in other subjects.

Another important feature of the data is that, for English and mathematics, the curves for GCSE grade B are much nearer to the curves for grade C than to those for grade A. In both mathematics and English, the one grade difference between a grade C and a grade B at GCSE equates to a single grade difference at A-level (although this is from grade E to grade D in mathematics and from grade D to grade C in English, for the reasons discussed in the paragraph above). However, increasing a GCSE score from a B to an A improves the median English A-level grade by a grade and a half, and the median mathematics grade by *two* grades. A-level can therefore be thought of as placing a 'magnifying glass' on the highest grades at GCSE (A and B), but not on the lower grades.

A third feature of the curves in Figure 1 is that the slopes of the curves for English are all very similar, as are those for mathematics, but the slopes for English are much greater than those for mathematics. This is related to the fact that the spread of A-level grades awarded to those obtaining a particular GCSE grade is much greater in the case of mathematics than English or of other subjects generally. In crude terms, GCSE predicts A-level performance better in English than in mathematics, which is, arguably, the reverse of what we might expect.

What is the nature of the gap?

In order to examine the nature of the gap between GCSE and A-level in mathematics, a group of eight teachers and researchers met for two days and looked at examination papers, marking schemes and candidates' scripts for both GCSE and A-level, for a range of examination boards. Only a limited sample of scripts could be made available given the short notice, and this considerably reduced the effectiveness of this strand, partly by reducing the sample size, and partly because there was no way of ascertaining the representativeness of the scripts. In some cases the papers came from an examination which had a very small entry; where there was pressure of time, these were the papers that were ignored. At both GCSE and A-level, the available scripts covered a wide range of attainment, although only Intermediate and Higher tiers were available at GCSE, and at A-level scripts around the middle grades (C, D and E) were over-represented.

The team was divided into three groups, which looked at questions relating to trigonometry, probability and algebra respectively. Each group read through all of the available scripts, noting questions which assessed material relevant to their focus, and looking at the kinds of responses made by candidates. The small size of the sample, and the questions over the representativeness of the scripts, require that great caution is exercised in interpreting the data generated. Furthermore, the considerable revisions to A-level courses to make them suitable for students who have achieved only National Curriculum level 7, implemented in 1996, makes drawing any suggestions for future policy very tentative. Nevertheless, some noteworthy themes do emerge from this analysis, which are discussed below, topic by topic. The reports below use the conventional labels used by teachers to describe different aspects of the mathematics in GCSE and A-level programmes such as 'algebra' or 'algebraic manipulation'. We should stress that our view of these 'topics' extends considerably beyond the narrow range of skills in these aspects of mathematics that are tested in GCSE and at A-level. However, given the focus of the current study, and the limitations on time noted above, we had to restrict our analysis to precisely those aspects of the topics tested at GCSE and A-level.

Trigonometry

At GCSE Intermediate level, all the trigonometry related only to right-angled triangles, while at Higher level, trigonometry included right-angled triangles, simple trigonometric equations, trigonometric graphs (where candidates had to find and plot values), sine and cosine rules, and 3-D problems. At A-level, trigonometry included right-angled triangle trigonometry, more complex trigonometric equations, sketching trigonometric graphs, identifying properties such as range, domain and periodicity in trigonometric graphs, sine and cosine rules, 3-D problems, addition formulae, trigonometric identities, trigonometry used in integration, and radian measure.

Trigonometry accounted for only 2-3% of the available marks in the Intermediate tier of GCSE, 10% in the Higher tier, and 20% at A-level, although these averages conceal considerable variation between syllabuses. In both the question papers and the students' responses, there was a discernible progression from GCSE Grade C to GCSE Grade A to A Level Grade N/E and to A Level Grade A/B, although there were some exceptions. For example, the sine and cosine rules, which A grade students at GCSE handled successfully, were only dealt with successfully by those who gained A-level grade D or better.

One worrying feature of the responses, particularly given the availability of graphic calculators, was the number of candidates at A-level who were able to manipulate trigonometry algebraically, but appeared to lack appropriate visual images which would have provided a fuller understanding.

Probability

There is little discernible gap between GCSE and A-level in probability. On the contrary, there is considerable overlap. Of the three topics studied, probability was the poorest predictor of overall grade, suggesting the possibility that ability in probability may be unrelated to ability in other areas of mathematics. The reasons for these findings are unclear, but it may be that comprehension of the context and problem-solving skills are more important than overall mathematical ability in solving probability problems.

Structured questions were answered far better at each level than unstructured ones, as might be expected. It is also the case that the type of questions set can be very predictable, often tackled by a process akin to 'colouring in by numbers', with the result that sometimes quite weak candidates scored highly on these questions. Another distinctive feature of probability is that almost no new mathematical concepts are introduced at A-level. The contexts are harder, and there is less supporting structure in the A-level questions, but there was at least one question at A-level which would not have been out of place in a GCSE examination. The difference between GCSE and A-level appears to be more in the maturity and comprehension required rather than in technical content. The A-level questions on conditional probability tended to be undertaken by the use of 'naive' strategies that would be expected at GCSE level, rather than by application of the more sophisticated methods that students could be expected to have encountered in their A-level studies.

Algebra

The clear message emerging from the scripts studied for this sample is that algebra is found disproportionately difficult by most candidates. Candidates scoring 50% overall hardly ever gained half of the marks available for algebra. Indeed, there were examples of candidates achieving grade A at GCSE without answering correctly (or even addressing!) any of the algebra items. Having said this, there was a fairly clear notion of progression in algebra. Students at grade C GCSE were able to draw and read values off quadratic graphs, and substitute positive integer values into algebraic expressions. Grade A candidates at GCSE could reliably substitute non-integral and negative values into expressions, change the subject of a formula in simpler cases, and multiply out pairs of brackets. At A-level, algebraic manipulation skills were very strongly related to overall grade. Furthermore, such errors as were made by those getting higher grades were almost entirely errors of negligence, whereas the errors of lower attainers also involved misunderstandings of fundamental principles. For example, candidates achieving higher grades might make errors of manipulation in expressing

$$\frac{2x+1}{x(2x-1)^2}$$

in partial fractions, while those candidates achieving the lower grades would attempt to find partial fractions of the form

$$\frac{A}{x}+\frac{B}{2x-1}$$

or

$$\frac{A}{x}+\frac{B}{(2x-1)^2}$$

rather than the correct

$$\frac{A}{x}+\frac{B}{2x-1}+\frac{C}{(2x-1)^2}.$$

Does it matter?

In order to validate the findings with regard to the size and nature of the gap, and also to examine the response of those responsible for admissions to A-level courses, interviews were conducted with a small sample of experienced tutors. Fifteen institutions were visited, ranging from the sixth form of a comprehensive school with some 24 students doing A-level mathematics in each of years 12 and 13 to a large further education college with some 200 A-level mathematics students in each year. The sample consisted of one teacher in a comprehensive school, two teachers in a school consortium, nine in sixth form colleges and four in further education colleges. The institutions were located in three geographical areas: Bristol, London and Manchester. In view of the small, and possibly unrepresentative, nature of the sample, it would not be wise to place too much reliance on the results of these interviews. Nevertheless, they do provide an interesting 'snapshot' of the views of a reasonable range of admissions tutors, which warrant further investigation.

The interviews were focused around eight pre-arranged questions, but the interviewees were encouraged to add any other comments which they wished to make. The responses have been grouped under the following headings:

Entry qualifications for A-level mathematics courses

Relation between entry and exit grades

Desired qualities of A-level students

Unsuccessful students

Algebraic skills of incoming students

Transfer from GCSE to A-level

Standards at A-level

Entry qualifications for A-level mathematics courses

All tutors interviewed would prefer incoming students to have a grade A or B achieved on the 'Higher' tier at GCSE. Most said that they would also take students who had achieved a grade C on the Higher tier, although these students are usually warned of likely difficulties with the course. Six tutors said they would also accept students with a grade B from the Intermediate tier, some with the caveat that they expect such students to have followed the Higher tier course for part of the time.

Others, however, will accept students with a grade C on the Intermediate tier, sometimes after consultation with the feeder schools. Some tutors emphasised that they give students who seem likely to get a grade C on the Intermediate tier a diagnostic test which is followed up with a workshop on algebra and a pack of materials to work through before commencing the A-level course. However, several tutors commented that they felt they should take students with a 'good grade' at Intermediate level, when they come from schools which enter pupils only for that level. It was suggested that the reason the schools offer only a limited range of levels is because they are concentrating on getting good results for their position in the 'league tables'.

Some tutors took the pragmatic view that, in fact, they teach anyone who is really keen to be on the course: "In the last resort, if they really want it, we always take them". Six tutors, three from FE and three from sixth-form colleges, said that they are sometimes pressurised by the senior management to take students whom they considered to be inadequately qualified.

Relation between entry and exit qualifications

Generally, the tutors we interviewed were reluctant to make any generalisation about a connection between entry qualifications and ultimate success at A-level, and although some quoted examples of students who arrived with grade As and enjoyed a very successful A-level course, most preferred to cite examples of exceptions: students with weak GCSEs who did well at A-level or those with good GCSEs who didn't.

Desired qualities of A-level students

Eight of the sixteen tutors interviewed mentioned specifically the need for students to have basic algebraic skills, although even if they lacked these basic skills, most interviewees felt that students could acquire quickly the necessary algebra if they were motivated to do so (although one was strongly of the view that if students did not arrive with basic algebra, they would never catch up).

However, in each case, it seemed that the students' attitudes to mathematics and to learning and learning skills were felt to be more important. One tutor made a special point of saying that GCSE coursework prepared students well to work independently and to do their own research.

Two tutors said that it is helpful if the student is studying at least one other related subject, such as physics. Students who choose 'Pure Mathematics and Statistics' often combine this with humanities subjects and were said not to be able to transfer any skills or knowledge between the two areas.

Several tutors said that the tendency of students to have part-time jobs interfered with their study, citing the "day-by-day need in mathematics to carry out follow-up work immediately after each session" as a particular problem. In other subjects, it was felt that students could catch up later if they had other commitments such as paid employment in the evenings, but in mathematics this is much more difficult. Two tutors mentioned the distraction of local discos and night-clubs as a hindrance to progress, particularly as it was the habit of such students to undertake paid jobs in order to afford them. One tutor said that this is characteristic of those doing single subject maths, and that "further maths students tend to be keen workers and have healthy interests such as climbing or family activities".

Unsuccessful Students

Most of the tutors identified a lack of algebraic skills as the main reason for a lack of success at the A-level course: "The weak student has not been exposed to (or certainly retained) basic algebraic skills and so finds it hard to pick up new ideas."

Despite the fact that GCSE performance was not *necessarily* regarded as a good predictor of A-level achievement as noted above, difficulties at A-level were frequently attributed to having done Intermediate level GCSE courses: "A typical weak student would come from an Intermediate GCSE and private tuition."

However, qualities such as lack of perseverance, lack of motivation, lack of ability to apply what they know, lack of confidence, reluctance to ask questions, and lack of effort were also deemed significant. In one case the problem was social, rather than mathematical, the student in question having had difficulty settling in to a new environment. Students who had been previously drilled into rule-based routines were also said to have difficulties with A-level work.

In almost all cases, tutors mentioned the support that they offer to students who encounter difficulties on the course, this help taking the form of specifically designed material, together with workshops and tutorials which usually take place in the tutors' non-teaching time.

Algebraic skills of incoming students

In order to find out what teachers of A-level mathematics expect of new students, interviewees were shown a set of algebra items from a GCSE Intermediate tier examination paper and another set from a Higher level paper and were asked how competent they would expect their incoming students to be with answering the questions.

Of the Intermediate level questions, most tutors felt that their incoming students would be confident in answering most of the items. The item that was regarded as doubtful was one that involved multiplying out brackets such as $(2x - 3)(x + 4)$; many tutors thought it unlikely that students could carry this out correctly, and several suggested that their students would probably not have met this type of question before.

None of the tutors would expect their students to be successful with all the Higher tier questions and some would not expect *any* of the Higher tier questions they were shown to be answered correctly. In the case of an item that asked for the expression $x^2 - 4$ to be factorised, three tutors suggested that students would not have encountered the difference of two squares before. Indeed, one tutor, who also teaches several GCSE classes, said that he would not bother to teach this process, as it took a fair amount of time to teach and, as the item carries very few marks on the examination paper, it was not time well spent.

An item concerned with sketching a function and interpreting simple transformations was thought to be more familiar to students, many of whom would use a graphical calculator with confidence. However, students were thought to be unlikely to be successful at an item involving tangents to a curve.

The lack of ability of students to do the Higher level questions was not, however, seen to be a problem since all tutors made it clear that this content is automatically taught again as part of the A-level course. A typical quote is:

> "Students couldn't cope with the questions cold—they may have seen most of them before but we re-teach it all anyway. Even if they did well at Higher level it is unlikely that they picked up many marks for algebra."

It should be noted that the fact that incoming students lacked facility with algebra was not, in general, seen as a problem. In their responses to other questions, particularly to the next one, on the gap between GCSE and A-level, almost all tutors said that any lack of algebra is dealt with as part of the A-level course. For example: "Algebra is not a problem, because we deal with it. Most of them acquire the necessary skills quite soon."

Transfer from GCSE to A-level

We now address the issue of whether tutors who teach A-level courses find a worrying gap between GCSE and A-level courses. Only three of the 16 tutors interviewed thought there is a 'gap' about which they have concern, one saying that if students had better algebraic skills they would manage the course better. All the remaining 13 tutors said they were not worried by a gap between GCSE and A-level, referring to the fact that the newer A-level text books address any lack of skills that the students might have. GCSE course work was said by several tutors to be a good foundation, but A-level didn't build on this adequately. Many students are apparently disappointed that some of the things they did at GCSE which they regard as interesting are not continued at A-level.

Standards at A-Level

All tutors acknowledged that the changes in content made it difficult to compare standards over the years, but, with two exceptions, tutors said that they considered that it was just as hard now to get a good grade at A-level as it was in the past, although they considered that it is now easier for students at the lower end of the range to get a pass. This was not necessarily seen as a disadvantage, several tutors saying that students who worked hard and were reasonably successful should achieve some rewards.

Two tutors spoke about the decline in the numbers of students taking double mathematics at A-level, which made comparison of standards very difficult. One said that there was no justification for comparing the standards of those taking single mathematics as no competent student ever took single mathematics 20 years ago: "In trying to compare them, you are looking at quite a different student with quite a different background of experience. They know more."

Discussion

The evidence presented in this paper does, we believe, present reasonably strong evidence that the 'gap' between GCSE and A-level in mathematics is greater than that for English for all but the very highest attaining students. The evidence also suggests that the gap is greater in mathematics than for the generality of other subjects, although the evidence here is less strong. It does seem to be a case of "everyone's out of step but mathematics", so that the suggestion made by Sir Ron Dearing in his review of 16-19 qualifications (Dearing, 1996a, p. 86), that comparability of standards should be achieved by a process of 'levelling up' seems rather perverse.

An analysis of a small sample of students' scripts at both GCSE and A-level indicated that the 'gap' such as it is, is most manifest in the area of algebraic manipulation - a finding supported by the interviews with admissions tutors for A-level courses. However, the interviews also reveal that these shortcomings in basic mathematical skills are by no means insuperable - indeed, it seems as if teachers of A-level have been overcoming these difficulties quite successfully for years.

The fact that weaknesses in facility with algebraic manipulation can apparently be rectified successfully by additional teaching within A-level courses contradicts assertions made by many commentators that these skills must be inculcated from an early age if they are to be acquired at all. Although it will need to be supported by more wide-ranging studies, the message coming from the research reported here is that there is no need to distort the GCSE curriculum, taken by almost all of the secondary school population, for the needs of the few who go on to A-level.

References

Dearing, R.: 1996a, *Review of qualifications for 16-19 year olds: appendices*, School Curriculum and Assessment Authority, London

Dearing, R.: 1996b, *Review of qualifications for 16-19 year olds: full report*, School Curriculum and Assessment Authority, London

Dearing, R.: 1996c, *Review of qualifications for 16-19 year olds: summary report*, School Curriculum and Assessment Authority, London

Wiliam, D. (ed.): 1996, *The step between GCSE and A-level in mathematics*, School Curriculum and Assessment Authority, London

THE NOVICE MATHEMATICIAN'S INQUIRY ABOUT NEW CONCEPTS: BESTOWING MEANING THROUGH AMBIVALENT USES OF GEOMETRICAL METAPHORS

Elena Nardi

University of Oxford

Abstract

The doctorate project on which this paper is based, is a study of the novice mathematician's conceptual and reasoning difficulties in their encounter with mathematical abstraction. For this purpose 20 Oxford first-year mathematics undergraduates were observed and audio-recorded in their weekly tutorials and interviewed twice in two academic terms. Learning Episodes were selected and arranged in terms of mathematical topics (Foundations of Analysis, Calculus, Topology, Linear Algebra and Group Theory). Data analysis aimed at the emergence of data-grounded theory and was balanced between the topical (epistemological) and cross-topical (psychological) perspectives. The final synthesis of the findings was arranged in terms of a number of themes that characterised the novices' learning behaviour. The Learning Episode presented in this paper exemplifies one of these themes - concept-image construction through the adoption of geometrical metaphors - *within the mathematical area of Group Theory and specifically the new concepts of* coset *and* equivalence class.

Introduction

Mathematics is defined to be an abstract way of thinking. Abstraction ranks among the least accessible mental activities. In an educational context the encounter with mathematical abstraction is the crucial step in the transition from informal school mathematics to the formalism of university mathematics. This transition is characterised by cognitive tensions. The study (Nardi, 1996a), from which the data and analysis presented in this paper were taken, aimed at t*he identification and exploration of the tensions in the novice mathematician's encounter with mathematical abstraction.*

Principles and Theoretical Background of the Study

As noted in previous presentations of parts of the study (Nardi, 1994, 1995, 1996a, 1996b, 1997) abstraction is meant both from a psychological perspective, i.e. that the advanced mathematics learner has to build up knowledge in an axiomatic way and learn how to reason deductively; and from an epistemological perspective, i.e. that the nature of the objects of advanced mathematical learning can extend beyond the physical or the numerical.

The structure and content of the data analysis reflect the juxtaposition and co-ordination of the following perspectives. The way the psychology of the individual learner is described is clearly influenced by:

- Piagetian ideas on how the transition to abstract forms of thinking takes place. Of particular importance are:
  - the notion of Reflective Abstraction, as specifically transformed for the needs of Advanced Mathematics by the theories developed within PME-AMT (the Advanced Mathematical Thinking Working Group of PME) (Dubinsky, 1991);
  - the theory of concept-image (Tall and Vinner, 1981; Vinner, 1991, 1983);
  - the notion of metaphor construction (Sfard, 1994; Sfard and Linchevski, 1994).
- Vygotskian (1986) ideas on recognising the strong interdependence between formal reasoning/thought and formal language.

The Lacanian psychoanalytical approach (1977) has influenced the part of the analytical discourse in this study with relation to the role of the unconscious in learning and its control by the learner as a means for conceptualising and overcoming Epistemological Obstacles (Sierpinska, 1994). The novice's induction to Mathematical Abstraction has been described as a process of conceptualisation and confrontation of Epistemological Obstacles as well as an enculturation process. The perspective on the novice's learning as an enculturation process was generally drawn from Hall's (1981) 'transition from informal to technical level' and Foucault's (1973) - from the 'rules of sense' to the 'rules of rationality' - archaeologies of knowledge. The perspective on this enculturation as a learning process built upon a didactical contract - the clarification of the conditions of this contract are described here as the ultimate task of the tutor-enculturator - is drawn from the theory of Guy Brousseau (1989).

In the above, learning is not seen as isolated in a cognitive vacuum but embedded in a sociocultural context. Therefore, the learner's cognition, while being personal and individually interesting, is also emphatically seen as taking place in a learning environment. In this case the context within which learning takes place is the Oxford undergraduate mathematics course. This study sought to construct a psychological profile of the novices' difficulties in their encounter with mathematical abstraction by probing into their expressions of learning. It is assumed here that cognition can only become visible and accessible through the learners' oral and written articulations of their mathematical thinking. As a thought process cognition is esoteric and inaccessible (Balacheff, 1990). In fact this is a phenomenological study of advanced mathematical cognition (Burrell and Morgan, 1979).

The Methodology of Data Collection and Data Analysis

Experience from a pilot study (Nardi, 1994) provided evidence that tutorials given to first-year mathematics students in Oxford can be a substantial source of data regarding the novices' expressions of mathematical cognition. These tutorials are 30-60 minute sessions in which a mathematics tutor and one or two first-year mathematics undergraduates discuss difficulties the students had with the problem-sheets given to them every week, based on the content of the lectures. Observation, audio-recording of the tutorials and interviewing of the students observed were chosen as the qualitative techniques through which access to these expressions would be achieved. Observation of tutorials (lasting 14 weeks, approximately 200 hours) on *Linear Algebra, Continuity-and-Differentiability*, *Topology, Sequences-and-Series* and *Groups-Rings-and-Fields* though relatively unsystematic was informed by the cognitive aims of the study as well as by research in the field of Advanced Mathematical Thinking. Two interviews were carried out with each student, minimally structured around mathematical topics that emerged, during observation, as particularly problematic for the learners. The openness of the selected methodological techniques was a natural consequence of the decision to ground any theory generated by this study on the data (Glaser and Srauss, 1967).

Tutorial and interview recordings were subsequently transcribed and tabulated in terms of their mathematical content and their didactical content. Via a gradually more selective process, a number of crucial Learning Episodes were extracted as the pivotal material for the analysis (Episodic Material). The rest, called Non-Episodic Material, was used as supportive material that enriched and contextualised the Episodes. Each of the Episodes is presented and analysed as a 'text' (Coulthard, 1985). Text here is the conglomeration of the recording, the transcript, the notes taken during observation and the contextual documents (e.g. problem sheets, lecture notes and reading lists).

The Episodic Material was then arranged in five sets of Analytical Texts in terms of mathematical content (*Foundations of Analysis*, *Calculus*, *Topology*, *Linear Algebra* and *Abstract Algebra*). The psychological observations on each Episode were collected and presented in a Topical Synthesis within each one of the mathematical topics. The Cross-Topical Synthesis of the theoretical abstractions of the study was based on the intermediate theorising that took place in the Topical Syntheses. Whereas the data analysis within each Episode and in the Topical Syntheses is balanced between the specific difficulties within each mathematical topic (micro-discourse on mathematical learning) and the cognitive patterns in the novices' learning behaviour, in the Cross-Topical Synthesis the aim was to describe overarching patterns.

In this paper I outline briefly one of these themes and exemplify it with one Paradigmatical Episode. Each Episode was presented as an Analytical Text. A number of Themes emerged in these Analytical Texts. These themes comprised the set of theoretical findings of the study. In the thesis (Nardi, 1996a) these themes were substantiated by a selection of Episodes, one or two Episodes for the sake of conciseness. These representative Episodes were called Paradigmatical Episodes. The content and analysis of the Episode presented in this paper also illustrate how inextricably the two perspectives of the analysis (topical/epistemological and cross-topical/psychological) are linked, that is how a piece of analysis, whilst totally immersed in a specific area of difficulties within a mathematical topic, emerges as a representative of an overarching theme.

An Example of a Theme Presented in the Cross-Topical Synthesis: the Ambivalence of Concept-Image Construction Through the Adoption of Geometrical Metaphors

There follows a brief outline of how some aspects of the novices'[1] concept-image construction was presented in the Cross-Topical Synthesis. Concept-image construction was identified in the study as the novices' predominant way of acquiring new concepts. The theoretical tool underlying the analysis was the Tall-Vinner theory informed by Sfard's theory on the accommodation of new concepts through the construction of metaphors. The topical and cross-topical syntheses are sets of theoretical abstractions generated by the analysis of the data collected while observing and interviewing the novices.

Acquisition of new concepts was an essential part of the students' enculturation into mathematical abstraction. Apart from, or in parallel with, interacting with the concept definition and building on previously established knowledge, the novices construct images of the new concepts by:

a) adopting metaphors that they seem to be familiar with, and

b) exploring the justification for the coming-into-existence of the new concepts.

Within the latter, the novices engage in concept-image construction by searching for the *raison d' être* of the new concepts. In the novice's economy of mind[2], a new concept justifies its coming into existence if it is of some utility. So, for instance, the students' prejudice for continuous functions may imply their view of limit as a redundant concept - what is the point of learning about $\lim_{x \to a} f(x)$ when simply calculating $f(a)$ gives the same result?

[1] The term 'novice' is used to refer specifically to the novice(s) participating in this study.
[2] cf the mathematician as *homo historicus*

Adoption of familiar metaphors, the former in the above-mentioned two ways the novices constructed images, was extensively exemplified in the students' reacting to the novelty of the abstractions in Vector Analysis: almost exclusively their action is embedded in the geometrical context of the line, the plane and the three-dimensional space. The excessively literal interpretation of this geometric metaphor, which was, I think, intended by the tutors as an illustration rather than a generic example, may be due to the fact that the students have associated the plane with convenient and familiar algorithms. The evidence in this study reinforces this explanation: the novices tend to draw on familiar modes of expression and reasoning when confronted with new situations – for instance, by trying to embed the new concept in their previous knowledge.

Finally, I note that substituting the power of the abstraction in the new concepts with the convenience of a familiar context and concentrating on a competent if narrow-minded execution of algorithmic tasks within this context, is more controversial than it actually sounds at first: indeed, specificity of context *is* a powerful initiator into new concepts; but maybe a too powerful one. In the Paradigmatical Episode presented below, the extensive adoption of a geometric metaphor in the context of Group Theory (regarding equivalence classes and cosets) seems to impede the learner's cognitive leap into the Abstract-Algebraic context.

An Example of a Paradigmatical Episode: Analysis (an Interpretive Account)

The transcript of this Episode is given as an Appendix to this paper.

The Analysis: An Interpretive Account

Literal Interpretation of a Drawing I: Equivalence Classes as Straight Lines: Fig.a is the tutor's visual representation of f, a mapping of G onto itself. In order to avoid a diagram in which G and Imf would be separated (a misleading idea since Im$f \subseteq G$), the tutor prefers to represent an element a of the domain as a dot and its equivalence class (generally defined as the set of elements in the domain that are mapped on the same value as a) as a line segment. This metaphorical representation however seems to escape Camille who interprets fig.a literally and wonders (C1) why equivalence classes are straight lines. T1 and Figure b are attempts to set the record straight and emphasise the metaphorical nature of the representation. Soon, however, Camille seems to repeat analogous interpretations with regard to the notion of coset (C7 onwards). C1 and C2 reveal Camille's preoccupation with the notion of an equivalence class which extends later, more intensely, to the relevant notion of a coset.

Literal Interpretation of a Drawing II: Cosets as Squares. A Multi-Faceted Tentative Construction of a Meaningful Image of the Concept of Coset: The discussion of the correspondence between the elements of a group and their equivalence classes evokes in Camille a query on another correspondence: 'the one-to-one correspondence between the conjugates of x and x'. Remarkably Camille demonstrates precise knowledge of the relevant definitions (centraliser, conjugate) as well as a relation between the two concepts. I note that, unlike Camille, most of the students at this stage in these tutorials were incapable of reproducing definitions of even simpler group-theoretical constructs mentioned in the lectures.

However, Camille in her demonstration of knowledge has not used the term *coset* at all. The term occurs for the first time in the tutor's words and captures Camille's attention. Subsequently it seems that she is preoccupied with the notion of *coset*: C3-C9 seem to be persistent, multiple attempts to imbue it with some meaning. C3 comes through as a surprisingly philosophical and abstract question which raises a very fundamental existential issue with regard to the notion of coset: what is surprising about C3 is that it comes in the middle of the tutor's describing a quite sophisticated construction (establishing a correspondence between the cosets of the centraliser and the conjugates of an element x in a group) and shifts the conversation from the strictly and specifically mathematical (represented by the tutor) to the metamathematical. Camille has been attentively listening to the tutor's demonstration of the construction and has given the very strong impression that, throughout, she has been processing the dense information provided by the tutor. C3 however illustrates that this processing must have been motivated mostly by the desire to construct a representation of coset - visual, *'material'* - than consume the tutor's argument. From then on, C3-C9 is a series of successive attempts at interpreting the concept of coset.

C3 is a quasi-platonistic enquiry on the nature of cosets as objects, as entities. (In fact, Camille is of British-Swiss origin and has been educated bilingually. *Matériellement* in French has the meaning of *actually, effectively, in a concrete and objective manner.*) Camille's in C3 do not necessarily act or interact. In C4 the questioning of the nature of these objects takes the form of an exploration of their *raison-d'-être*. C5 is a dissection of a coset which equates a coset with how it comes into existence. I note that so far T2-T4 do not seem to have a direct impact on the genesis of Camille's ideas of what a *coset* is. C6 is a geometrical interpretation of C5 perhaps derived from the notion (and notation for) transformations, and in particular translations. The tutor carefully tunes in (T5) but Camille accelerates her tentative condensation of her conception of coset in a geometrical image in a questionable way: C7 (in parallel with C1) illustrate how the line between a metaphorical and a literal interpretation of a picture is thin and severely disguised under the heavy-weight impact of visual imagery. The tutor is surprised and alarmed (T6) by Camille's intention to '*apply* [this idea] *on squares*'. C8 is evidence that Camille is too preoccupied with her image construction to be influenced by T6 and she furthers the interpretation of her Figure c in a less controversial but highly ambiguous way. T7 is one more effort on the tutor's part to tune in and transform the student's images from within. Surprisingly then Camille shifts to a more abstract property of cosets (that they are '*of the same size as the original*' in C9) in which however the geometrical jargon ('*size*') is maintained. The tutor (T8) has completely adopted Camille's metaphor and contributes another observation on cosets, that they partition the group.

Camille ceases the effort to interpret further the notion of *coset* once she acquires an image of *cosets* that is satisfying and clear to her. That Camille is content with what she has acquired can be assumed on the basis of the evidence, given during the six months of observing this student, that she does not bring a conversation to an end until she acquires a satisfactory (to her) understanding. However an issue that C3-C9 raises is whether the quality of the acquired perception of a coset - via a multiplicity of metaphors and visual representations - justifies Camille's eventual sense of satisfaction. Given that the tutor cautiously surrenders in adopting Camille's metaphor but does not cross-check whether the intended (by the tutor) and the acquired (by Camille) image of a *coset* coincide, the questions raised by this issue ought to remain open.

Summary of the Episode and Conclusion

In the above, a student, who exhibits a remarkable knowledge of the definitions of the concepts involved in the discussion, is engaging in a meaning-bestowing process with regard to the notions of equivalence class and of coset. The student asks the tutor about the *raison-d'-être* of the concepts and her efforts are characterised by a tendency to use metaphors of some regular geometrical shapes in order to construct a mental representation of the concepts (equivalence classes as *straight lines*, cosets as *squares*). Evidence was given in the Episode that these geometrical representations are interpreted literally by the student and this raised the issue of a potential cognitive danger built into the use of geometrical metaphors. The tutor here demonstrated a considerable flexibility in thinking in terms of the student's metaphors (actually it is the tutor who sparks off the use of geometrical representation in this tutorial) but in the end there doesn't seem to exist any guarantee that the didactical use of metaphorical discourse has resulted in the tutor's intended concept image of the notion of coset.

Embedding the Episode into the Theorising Process of the Study and a Potential Extension of the Study

As explained earlier, the Episode and its analysis given above, are an example of an Analytical Text. The observations in the analysis about the learner's behaviour informed the Topical Synthesis on Group Theory, which in turn, along with other evidence from the tutorials on the use of geometric metaphors as a means to concept-image construction, has informed the Cross-Topical synthesis.

The Episode, and its analysis, come to an end remaining inconclusive about the didactical implications of the novice's extensive use of various geometrical metaphors in her attempt to construct concept images of cosets and equivalence classes. This didactical inconclusiveness was typical in this study, given that the focus was on mapping out the novice's learning behaviour and not addressing issues of teaching - even though it was inevitable, for instance in this Episode, to hint at the ambivalent effectiveness of relying excessively on the unclarified use (Brousseau, 1989) of metaphorical discourse, visual or verbal. As mentioned briefly in the introduction, the novice's encounter with mathematical abstraction was treated in the analysis as an enculturation process (Hall, 1981) and as such the role of the tutor-enculturator emerged as a pivotal determinant of the learner's behaviour. In fact in most cases the enculturation process turned out to be such a complex interactive process that disentangling the learning from the teaching aspects was extremely difficult. This strong interdependence between teaching and learning indicated a new direction as a potential extension of the study: a shift of focus towards the interaction between the learner and the teacher rather than focusing strictly on learning.

At this stage a first step towards this shift of focus is taking place as part of the feedback to the participants of the study and as the launch of a new enterprise - the tranformation of the psychological findings of this study into pedagogical practice. The tutors have been approached again and asked, in a series of interviews, to reflect and comment upon the data and the analysis of the study. Their responses to the extracts and interpretive accounts will trigger off a new study, with more direct pedagogical aims, possibly designed on the basis of a collaborative tutor-researcher intervention into the tutoring process.

References

Balacheff, N.: 1990, 'Beyond a psychological approach: the psychology of mathematics education', *For the Learning of Mathematics* **10**(3), 2-8.

Brousseau, G.: 1989, 'Le contrat cidactique: Le Milieu', *Recherches en Didactique des Mathématiques* **9**(3), 309-336.

Burrell, G. and Morgan, G.: 1979, *Sociological Paradigms and Organizational Analysis*, Heinemann, London.

Coulthard, M.: 1985, *An Introduction to Discourse Analysis*, Longman, London.

Dubinsky, E.: 1991, 'Reflective abstraction in advanced mathematical thinking' in Tall, D. (ed.): *Advanced Mathematical Thinking*, Kluwer Academic Publishers, Dordrecht / Boston / London, pp. 95-123.

Foucault, M.: 1973, *The Order of Things: An Archaeology of Human Sciences*, Vintage Books, New York.

Glaser, B. G. and Strauss, A. L.: 1967, *The Discovery of Grounded Theory: Strategies for Qualitative Research*, Aldine de Gruyter, New York.

Hall, E. T.: 1981/1959, *The Silent Language*, Anchor Press, Doubleday, New York.

Lacan, J.: 1977, *Écrits: A Selection*, Routledge, London.

Nardi, E.: 1994, 'Pathological cases of mathematical understanding' in *Proceedings of the 18th International Conference for the Psychology of Mathematics Education*, Lisbon, Portugal, Volume III, pp. 336-343.

Nardi, E.: 1995, 'The novice mathematician's encounter with mathematical abstraction: the case of accumulation point', *European Research Conference on the Psychology of Mathematics Education*, Osnabrück, Germany, pp. 54-57.

Nardi, E.: 1996a, *The Novice Mathematician's Encounter With Mathematical Abstraction: Tensions in Concept-Image Construction and Formalisation,* Unpublished doctoral thesis, University of Oxford, Oxford.

Nardi E.: 1996b, 'Tensions in the novice mathematician's induction to mathematical abstraction', in *Proceedings of the 20th International Conference for the Psychology of Mathematics Education*, Valencia, Spain, Volume IV, pp. 51-57.

Nardi, E.: 1997, 'The novice mathematician's encounter with mathematical abstraction: a concept image of spanning sets in vectorial analysis', *Educacion Matematica,* **9**(1), Grupo Editorial Iberoamerica, Mexico.

Sfard, A.: 1994, 'Reification as the birth of metaphor', *For the Learning of Mathematics*, **14**(1), 44-55.

Sfard, A. and Linchevski, L.: 1994, 'The gains and the pitfalls of reification: the case of algebra', *Educational Studies in Mathematics* **26**, 191-228.

Sierpinska, A.: 1994, *Understanding in Mathematics*, The Falmer Press, London

Tall, D. and Vinner, S.: 1981, 'Concept image and concept definition in mathematics with particular reference to limits and continuity', *Educational Studies in Mathematics* **12**, 151-169.

Vinner, S.: 1983, 'Concept definition, concept image and the notion of function', *International Journal of Mathematics Education in Science and Technology* **14**(3), 293-305.

Vinner, S.: 1991, 'The role of definitions in the teaching and learning of mathematics', in Tall, D. (ed.), *Advanced Mathematical Thinking*, Kluwer Academic Publishers, Dordrecht, pp. 65-81.

Vygotsky, L. S.: 1986, *Thought and Language*, MIT Press, Cambridge, MA.

APPENDIX: A Paradigmatical Episode

Context: This is the beginning of the thirty-minute weekly tutorial with student Camille. It is nearly the end of Hilary Term (end of February - beginning of March). The italicised parts of the text are the actual words used by the tutor and the student.

Structure of the Episode: In the following, Camille asks the tutor about the ideas she has found problematic in the lectures: the proof of one of the Isomorphism Theorems and then the definitions of centraliser and conjugate. Underlying her questions seems to be her difficulty with the new notions of equivalence class and, most severely, with the notion of coset.

The Episode:

Camille says she has been confused by *'what the lecturer referred to as something like a~b when f(a)=f(b)'*. The tutor replies that the lecturer might have been trying to prove one of the Isomorphism Theorems for Groups:

> if $f{:}G \rightarrow G$, where G is a group, then there is a bijection between the equivalence classes of the equivalence relation defined by '$a \sim b$ when $f(a)=f(b)$' and Imf.

The tutor draws Figure a to illustrate that if b and b' are different then $f^{-1}(b)$ and $f^{-1}(b')$ are different. Camille looks at Figure a and asks:

C1: *Why are they all straight lines?*

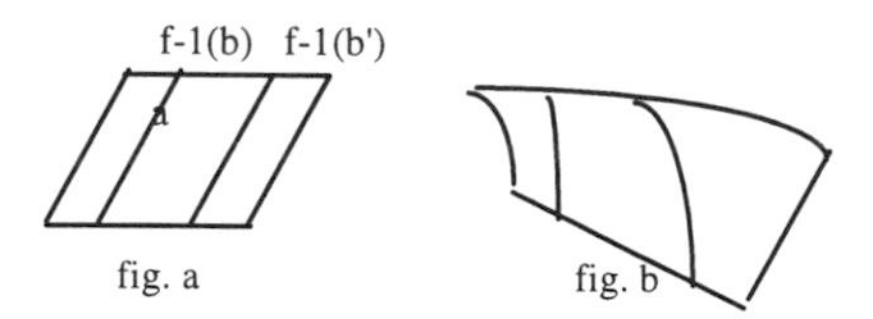

T1: *Because I drew them this way. It doesn't mean anything. It's only the way I drew it. And you can draw little 'squiggles' if you like* (Figure b).

C2: *And these* [the 'squiggles'] *are the equivalence classes?*

The tutor nods in agreement: the set of all elements equivalent to b is the equivalence class of b and mapping b to its equivalence class establishes a correspondence between the elements of the group and their equivalence classes. In the ensuing discussion [not presented here because it is not directly relevant to the theme of this paper] Camille seems to be sceptical about whether the tutor's and the lecturer's way of presenting this correspondence are equivalent. At the end of this part of the discussion she sounds convinced and moves on to her next question: can the tutor prove that *'there is a 1-1 correspondence between the conjugates of x and x?'*. The tutor asks Camille to define $C(x)$ and Camille defines the centraliser of x as

$$C(x) = \{h \in G\textbf{:}\ xh=hx\},$$

the conjugate of x as

$$\text{'}g^{-1}xg \text{ for some } g\text{'}.$$

and also points out that

$$\text{if } g \in C(x) \text{ then } g^{-1}xg=x.$$

So, says the tutor, the size of $C(x)$ tells you '*how many repetitions you get if you try to write x as a conjugate*'. So, she continues, if two elements g_1 and g_2 give the same conjugate they belong to the same coset of $C(x)$. And vice versa. This happens only when $g_2g_1^{-1}$ belongs to $C(x)$. Therefore there is a 1-1 relation between cosets of the centraliser and conjugates of x and these have the same number of elements. Camille is quiet and looks sceptical. Then she asks:

C3: *What are cosets materially?*

T2: *What do you mean by that?*

C4: *If we have a group G and a subgroup H why do we bother to find the cosets?*

T3: *Because of results like this. They turn up naturally.*

C5: *Cosets are a group multiplied by an element in the big group.*

T4: [hesitantly] *Yes...it's a set...*

C6: *Cosets are just a moving...*

T5: *That's right. That's one way... you can look at it as translates of a subgroup... sort of multiplying g with everything in H and it shifts it...*

C7: [after a pause] *So if we have a square of size one and then the group G is like this...*(fig.c)

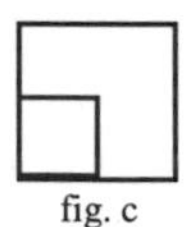

fig. c

T6: *You have to be slightly careful... It is slightly... don't think about in ... you're not thinking of applying it on squares, are you?*

C8: *So then it would have four cosets?*

T7: *Mmm... if the subgroup was a quarter of the size of the whole thing, yes... it would have four cosets... that's right.*

C9: *And the cosets are always the same size as the original.*

T8: *That's right. As we know they partition the group.*

The tutorial continues and closes with discussions on other mathematical topics.

'FLIPPING THE COIN': MODELS FOR SOCIAL JUSTICE AND THE MATHEMATICS CLASSROOM

Tony Cotton

University of Nottingham

Abstract

This paper offers a rationale for the development of a theory of social justice to support research in mathematics education. A Rawlsian perspective (Rawls, 1971) is taken as a starting point. This perspective is developed using the work of Iris Marion Young (1990), Cameron McCarthy (1990) and Godfrey Brandt (1986) to offer arenas for research in mathematics classrooms by exploring injustices within schools and within classrooms. The paper ends by calling for the focus for such social research to be transformational as opposed to explanatory.

> Justice, like many other moral attributes, is best defined by its opposite.
>
> (J. S. Mill, in Ryan, 1993, p. 52)

Introduction

Several locations inspired this paper so I will attempt to start in all of them. The first is a corridor in Valencia University following the presentation of a paper exploring issues of social research in mathematics education (Cotton and Gates, 1996). A colleague, who had chaired the session asked, "So, what is social justice? Is it simply a matter of flipping the coin?". She suggested that social justice could be read as the flip side of injustice. This idea of flipping a coin struck me as a good title for a paper - so here it is!

The second location is a bookshop near the Institute of Education in London. In 1994, I presented a paper at a BSRLM day conference exploring a journey to research issues of social justice in a socially just way - still a vexing problem (Cotton, 1994). Over lunch I bought three books. This paper grew from the ideas in two of these books: *Justice and the Politics of Difference* by Iris Marion Young (1990) and *Realising Rawls* by Thomas Pogge (1989). The first offered me a model I developed to explore social justice in mathematics classrooms and the second supported my use of a Rawlsian concept of social justice.

The third and most recent location for my introduction is the Symposium on Social Justice at the British Educational Research Association (BERA) Conference in Lancaster during 1996. Many papers were at best only tangentially related to social justice, and those models of social justice presented offered little to research in mathematics classrooms. It seemed appropriate, therefore, to work on this paper.

A word about the style of this paper. My decision to offer the paper as two parallel texts is influenced by an event in summer 1996. I attended a lecture entirely in Spanish, with overheads in English. I speak very little Spanish yet found the session one of the most stimulating I have ever attended. I felt liberated from having to listen to and interpret/understand every word spoken - I could choose to focus on either the overheads or a snippet of the lecture I had 'translated'. This allowed me to concentrate on my interpretation(s) of the text rather than merely interpreting the presenter's words. All this secure in the knowledge that I could read the text later to explore issues in greater depth. So the paper you have in front of you is a script: the right hand column offers a copy of the transparencies I used when presenting this paper at the BSRLM day conference in November 1996, the left hand column the spoken text. The style tries to mimic the thinking spaces we necessarily insert into presentations; those moments when we focus our thoughts either on an overhead transparency or an idea which has come through the presentation, and briefly absent ourselves from the presenter's presence. The physical spaces on the page can be seen as an invitation to interact with the text in a physical way through the reader adding their own jottings, notes, questions or ideas. The style also attempts to recognise the multiple texts/experiences on which we draw to create what we may call knowledge.

Reading is an exercise in listening.

(Lyotard, 1992, p. 117)

Discussions during a 1996 BERA symposium suggested that 'social justice' as a concept appeared when the terms 'anti-racism' and 'anti-sexism' had reached the end of their shelf-life. I feel very uncomfortable with that view. For me the term became important as I realised that activity I was engaged in under the guise of anti-racism drew heavily on ideas I had examined in my classroom when focusing on the underachievement of girls or the fact that the bottom sets I taught seemed to consist entirely of the 'kids from the tower block'. It became too simplistic to see issues simply in terms of gender, race or class - there were clearly large areas of common ground, as well as particular issues which affected particular definable groups.

I was also becoming less certain as to how I might measure success in these areas - if I reached the stage in school where my A-level group had 50% girls and 50% boys in it, or where my bottom set reflected the catchment area of the school accurately, was this success, or was I simply substituting an alternative set of winners and losers with no net change to the way in which school, or (more importantly) the pupil's world outside school operated. In retrospect it appears as though I was attempting to mould the nature of the learners in my classrooms to match a fixed view of mathematics and the teaching/learning of mathematics, not, as I would now suggest is important, attempting to explore possibilities for structural change.

"The entire history of social improvement has been of a series of transitions, by which one custom or institution after another, from being of supposed primary necessity of social existence, has passed into the rank of a universally stigmatised injustice and tyranny. So it has been with the distinctions of slaves and freemen, nobles and serfs, patricians and plebeians; and so it will be, and in part already is with the aristocracies of colour, race and sex."

(John Stuart Mill, quoted in Ryan, 1993, p 71).

Later I found myself almost literally moving 'allegiances' as I became an advisory teacher in multicultural and anti-racist education. Gender was off my agenda, I was now a 'specialist' in issues of 'race and maths education'. It became difficult to create alliances across these divides of oppressions, a particularly strange world in which to operate as a white, middle-class male.

"(there is a) tendency to play off sexism, class, disability and racism" (and view) "these systems of oppression alongside each other in competitive relations within a context of marginality."

(Brandt, 1986, p. 69).

It seemed to me that the biggest influence on me as a teacher, and on the beliefs I hold about educating in general and about mathematics education in particular, is that which greatly influenced my work as an 'anti-racist' teacher in schools. Similarly, all the work I engaged in under the banner of multicultural/anti-racist education had a place in exploring issues in mathematics classrooms which affect all groups disadvantaged by the mathematics education they are offered. I required a way to view my work as having a coherence across issues. A view of social justice offered by Bennison (1984) helped me. She suggests we should treat people the same in the ways in which they are the same and differently in the ways in which they are different. For me the term social justice defined a way of working which accounts for and works with the links between oppressions, inequalities and exploitations which we see inside and outside our schools and classrooms.

"no understanding of racism in society or in any of its social institutions, such as education, can be attained without employing a theoretical framework which explicitly and structurally recognises and accounts for the connections between the oppressions, exploitations and inequalities associated with the notions of race, class and gender."

(Chris Mullard, quoted in Brandt, 1986, pviii).

Unfortunately the term Social Justice has become capitalised - it is now something which can be linked to particular political parties and to factions within political parties. In many ways the term has become devalued. The final bullet point in the Commission on Social Justice Report (1994) suggests social justice will not be fought for by 'New Labour'. The phrase 'where possible' offers little hope of commitment to structural change. The concept of 'unjustified inequalities' reflects the Victorian view of the 'feckless poor'. We require a clarity in the educational view of social justice - if only to challenge what could become the trivialisation of the idea.

Social justice is defined by the Report from the Commission on Social Justice as a hierarchy of four ideas :

- *The equal worth of all citizens.*
- *Each citizen's equal right to meet their basic needs of food, shelter and other necessities.*
- *The need to spread opportunities and life chances as widely as possible.*
- *The need to eliminate where possible unjustified inequalities.*

(Commission on Social Justice, 1994, p. 1)

For me social justice represents a shift in thinking from the stance of 'equality' or equal opportunities. Equality can seem to suggest a norm towards which we should strive. It does not easily accept and value difference. Attempts have been made to address this issue through slogans such as 'equal but different'. Maybe social justice can be seen as the beginning of a theory around this slogan.

"Justice is not equality, because equality is concerned only with results, and not how they were arrived at, and equality is only concerned that people should be treated the same, whereas justice is concerned to consider each case on its own merits, treating, if necessary, different people differently."

(Lucas, 1980, p. 4).

A view of justice which does take account of difference can be seen in the legal use of the term. An interview on Radio Four's 'Today' programme showed how the view of justice as the ability to treat people in different ways depending on the 'merits' of the case can be at odds with present political realities.

Lord Donaldson argued that judges should maintain the right to weigh up the individual circumstances of each case before sentencing. It would not always be appropriate, he argued, to treat each case equally. Later in the discussion Michael Howard used the term 'justice' quite differently - in his argument 'justice' was something which was dispensed; he used the phrase "we want to bring persistent offenders to justice", in other words send them to jail. I had visions of an Orwellian prison with 'Welcome to Justice' in neon lights over the front gate.

For me Social Justice is also to do with power. It is do with how individuals and groups of individuals can feel powerful or be made to feel powerless. It is to do with individuals feeling in control of decisions which affect the way they live their lives. We feel as though an injustice has been done when someone takes a decision which affects us personally or emotionally and with which we disagree but against which we have no power to argue. Social Justice in this sense is linked to the power to make life choices without being denied access to particular life chances through discriminatory practices. Social Justice is also having the power, as well as the right, to fight practices we perceive as unjust.

You will never understand,

How it feels to live your life,

With no feeling of control,

And with nowhere left to go.

(Pulp[1] - Common People).

[1] A popular British band of the 1990s

Where does mathematics fit into all this? As a child of eleven, I was taught in the 'top' group in my Primary school. On Fridays we had a mental arithmetic test. The child who scored highest in the test sat in desk 1 (next to the teacher) for the following week, the child 'coming second' sat in the next desk and so on. After a week or two I realised that by gettingtwo or three questions wrong I would be about fifth in the test and get to sit by the door. I could engineer this result as I was confident that I knew all the correct answers and so could deliberately make my two or three errors.

I offer this as a metaphor for confidence and skill in mathematics carrying with it power over life choices. A teacher on a course brought this memory to the front of my mind when she told of a similar experience, although hers had left her feeling powerless and out of control. She felt she could not 'do' mathematics. Fridays brought with them anxiety and panic. She knew she would do badly in the test, would be made to sit in a 'lower' desk and worst of all that her mother would see she had failed when she collected her after school. I also note the gender of the tellers of the two stories.

The issue of 'power' is becoming increasingly important in our research community - I cannot even begin the debate here. My view of power is more essentialist than a Foucauldian view as it draws heavily on the work of Rawls - who feels able to define individuals as 'least powerful' in certain institutions. A view which I have sympathy with.

Rawls and Social Justice

Although Rawl's work has been criticised heavily for its romanticism, for its detachment from historical and political realities, and for its neglect of the factors of sex, race and class, it offered me a useful starting point for my exploration of social justice and education. It also has allowed me to look at ways in which institutions of education can be examined in order to move towards 'more just' ways of working. It opens up possibilities for change.

" most important upshot of Rawl's work - a broad debate about the justice of existing institutions and feasible avenues of institutional reform. Indeed Rawl's work has a unique affinity for a concrete debate about justice because of its commitment to the primacy of the practical."

(Pogge, 1989, p. 3).

Rawls uses the metaphor of 'halving an apple' to explain the basis of his theory. If two people are sharing an apple, one person cuts the apple and the other has the choice of which half they want. The first person will be as fair as possible in cutting the apple in order to ensure they receive a 'fair' share. The creation of his socially just Utopia takes place through a discussion by a group of people shrouded by a veil of ignorance - they do not know whether they are male or female, what their ethnic background might be, what their family situation is or what their historical backgrounds are.

"justice as fairness ... does not look behind the use which persons make of the rights and opportunities available to them in order to measure, much less, to maximise, the satisfaction they achieve. Nor does it try to evaluate the relative merits of different conceptions of the good. Instead it is assumed that members of society are rational persons able to adjust their conceptions of the good to their situation."

(Rawls, 1971, p. 94).

Rawls suggests, taking perhaps a rather pessimistic view of human nature, that this group would seek to protect themselves from harm in the new society they were creating. To build a just society, we should create a society as if our enemy could choose the position in which we are placed within that society.

"Social institutions are only just if they can be defended to each of their members on the basis of the contribution they make to his (sic) good as assessed from his (sic) point of view."

(Daniels, 1975, p. 172).

Perhaps Rawls offers us ways to critically examine our institutions and our classrooms. Can we apply his tests of justice? Do decisions we take as to the arrangements within our classrooms and our institutions always benefit the 'worst off' amongst our learners? Would we feel comfortable if we thought that our enemies could decide where to place us, or our own children, in order for us to learn mathematics within our schools.

If the answer to either of these questions is no, in what ways would we alter what we teach, or the way that we teach it to accord to Rawlsian justice?

I am reminded of a recent conversation with a group of 11 and 12 year olds just starting at secondary school. I had worked with them extensively at their primary school and was visiting them to talk about how useful they had found the mathematics we explored in the primary school now they were at secondary school. I happened to arrive on the day in which they had been placed in sets according to 'ability'. Two of the pupils were adamant that the new school wasn't fair.

They had been placed in the bottom set and complained bitterly that the mathematics was repetitive and too easy. A social justice perspective would suggest the school look carefully at this process as it is clearly not seen by these pupils to be to their best advantage. Indeed a Rawlsian interpretation of social justice only allows 'unfairness' within an institution if the 'unfairness' benefits the worst off within that institution.

Another measure of social justice put forward by Rawls is the right of all people to decide on a rational life plan which is designed to permit "*the harmonious satisfactions of his (sic) interests*" (Rawls, 1971, p. 93). A rational life plan is one which "*cannot be improved upon, there is no other plan which, everything taken into account, would be preferable.*" (Ibid. p. 93). Education in general and mathematics education in particular is clearly important in bringing this 'rational life plan' to fruition. Academic mathematics qualifications are used as a filter to future career prospects and any 'rational life plan' involving moving into a profession must include success in mathematics. However there are clearly many injustices within the assessment structure in our schools, even were these to be eradicated we have already seen that equal qualifications do not mean equal access for all to future career choices.

"Today, people's life chances are even more powerfully affected by their education than in the past. Not only are people without qualifications far more likely to out of work, but the earnings gap between people with a university degree or its equivalent, and those without, is rapidly growing wider."

(Commision on Social Justice, 1994, p. 42).

Rather than suggest that all must be given an equal opportunity to succeed academically at mathematics within the present structure, this suggests to me an arena of research to be that of the assessment processes used by mathematics educators in our schools. How can this be transformed to meet the conditions of social justice?

Rawls also suggests that within a socially just society, individuals and groups would feel able to participate actively in the democratic organisations within that society. This view of social justice raises many questions with regards to the mathematical education we offer in our schools. In what ways have the curriculum we adopt and the pedagogies we prefer addressed the issue of which skills and ideas mathematics learners in our classrooms will find useful to become fully functioning members of a democratic society.

"All citizens should have the means to be informed about political issues. They should be in a position to assess how proposals affect their well being and which policies advance their conception of the public good. Moreover, they should have a fair chance to add alternative proposals to the agenda for political discussion."

(Rawls, 1971, p 225).

Have we begun to address the idea of 'mathemacy' as Ole Skovmose (1994) calls it. That is that we demand not only a skills base, but more than that, a knowledge of how we may employ these mathematical skills, and a reflective knowledge which allows us to understand how our mathematical choices affect the ways in which we view and create our worlds through the results of our mathematising.

"the growing demand from previously excluded groups - women, ethnic minorities, disabled people - for a political system that includes them and better reflects their concerns and demands."

(Commission on Social Justice, 1994, p. 86).

This view of 'mathemacy' envisages the possibility of constructing a learning of mathematics which is designed to support learners in their development as reflective adults capable of using their mathematics to critique and challenge structures within society. If the purpose of teaching mathematics was to enable our learners to construct a better society we would clearly have to re-evaluate our notions of curriculum and pedagogy.

"rooted in the spirit of critique and the project of possibility that enables people to participate in the understanding and transformation of their society."

(Skovmose, 1994, p. 27).

It would seem to me that these ideas are entirely compatible with the desire that mathematics classrooms should be places where we educate both for mathematics and for a society based on ideas of social justice. It also suggests an acknowledgement that mathematics rather than being a tool to be used to interpret and explain the world around us is also used to create our world. I think this is an important point. We cannot move towards a mathematics for justice without questioning our notions of the nature of mathematics and the nature of mathematical knowledge. Indeed by exploring the social perspectives of mathematics education we begin to question many of the unjust practices present in our schools today and the search for alternatives begins.

"our approach to solving the social problems of mathematics and the problem of mathematics as a social problem will necessarily focus on social roles and institutions. New social circumstances and arrangements will give rise to new conceptions and forms of mathematics. We cannot anticipate these new conceptions and forms; to a large extent we cannot even imagine them. We can imagine practising, teaching, producing and using mathematics in new ways. This does not require attacking all social ills at all levels simultaneously; it does, at the very least, require that we approach, revisions, reforms, and revolutions in mathematics always with an awareness of the web of roles, institutions, interests and values mathematics is embedded in and embodies."

(Restivo, 1993, p. 276).

However, although this Rawlsian perspective offers useful models on which to build a theory of social justice and mathematics education it has been critiqued as viewing the values of justice and autonomy as moral issues detached from everyday human behaviour. This narrow view of social justice is challenged (Gilligan, 1982, 1988; Collins, 1990) by the assertion that for many women, the notion of care is a key to the way that moral decisions are made.

The push for autonomy within a society leads to a detached view of an individual, living within a hierarchically ordered society, whereas the values of care and attachment create a world of individuals within an attached network of relationships.

Incorporating the idea of 'care' within a social justice framework offers extra possibilities for transformation rather than adaptation, and again moves on from ideas of equality as equal, turning towards social justice as a transforming power. Gilligan's work has been criticised (Ernest, 1989; Collins, 1990) for relying entirely on middle class, white women within its sample. However, by attempting to pull together the links between the inequalities, exploitations and injustices suffered by different groups we begin to move towards a coherence which allows us to operate as researchers.

A new model of social justice must include ideas of care and connection with our family and cultural roots if it is to be a useful model. So mathematics for social justice must include a perspective of care.

"Carol Gilligan's work suggests that there is a female model for moral development whereby women are more inclined to link morality to responsibility, relationships, and the ability to maintain social ties. If this is the case, then African-American women again experience a convergence of values from Afrocentric and female institutions."

(Collins, 1990, p. 219).

We must not strive simply to produce autonomous, independent human beings, ready to play an aggressive role in pushing forward the domestic economy, or confident to take their place fighting for a place in a new job market, but must also look towards a pedagogy in mathematics which encourages values of sharing, co-operation, joint labour and skill sharing. Most importantly we must involve multiple perspectives when viewing actions and interactions in our classrooms: we must acknowledge difference rather than foster homogeneity. A colleague who works with bilingual learners, with language as her usual focus, noted the difference between her approach to responding to a piece of writing and a piece of mathematics. She described how, when commenting on a story, she always responds personally to the context before remarking on technical matters. She describes her marking of mathematics as "a series of ticks and crosses which children use to position themselves on a success/failure scale."

If we are to introduce the idea of 'care' and 'connection' to our mathematics teaching we clearly must offer contexts within which individuals can identify themselves and their colleagues and to which they can respond personally.

"the feminist potential of Rawl's method of thinking and his conclusions is considerable. The original position, with the veil of ignorance hiding from its participants their sex as well as their other powerful characteristics, their talents, circumstances and aims, is a powerful concept for challenging the gender structure."

(Okin, 1991, p. 196).

Exploring Injustice in Schools

As a researcher interested in exploring issues of justice in mathematics classrooms through looking at injustices presently built into our practices I required arenas in which to explore and act . Godfrey Brandt's study *Towards an anti-racist pedagogy* (Brandt, 1986) offered me three such arenas to which I added assessment. These arenas can be summarised through the following questions; What do we teach? How do we teach? What do we value as knowledge and ability? How do we and the learners in our care feel about our practices and our institutions?

Injustice is perpetuated through:

- *Mathematics Syllabi*
- *Mathematics Pedagogy*
- *Assessment methods in Mathematics*
- *The social and cultural environment within which we learn and teach mathematics*

(After Godfrey Brandt).

Cameron McCarthy (1990) offers four relationships which would suggest injustices were present in classrooms. A competitiveness which leads to individuals or groups becoming isolated from mathematics and from mathematics learning: this competition can be seen as a competition for access to education, a competition for credentials from education, as well as the competition for the scarcity of resources and teacher time.

Injustices are played out in the mathematics classroom through relations of:

- *Competition*
- *Domination*
- *Exploitation*
- *Cultural selection*

(after Cameron McCarthy)

The relationship of domination of one group over another in the classroom, of one teaching style over another, of teacher time, and of resources clearly takes us back to the exploration of power relations. Exploitation in schools is evidenced in schooling as preparation of individuals for 'appropriate' life plans rather than the rational life plans of Rawls. Finally the relationship of cultural selection has echoes of Bourdieu's (Bourdieu and Passeron, 1977) ideas of cultural capital.

Iris Marion Young (1990) offers five faces of oppression. The experiences of exploitation, marginalisation, powerlessness, cultural imperialism, and violence are shared by all individuals belonging to oppressed groups at one time or another. These oppressive forces do not exist separately but can be used to observe instances of injustice within mathematics classrooms.

Injustices can be seen in individual experiences of:

- *Powerlessness*
- *Violence*
- *Exploitation*
- *Marginalisation*
- *Cultural imperialism*

(after Iris Marion Young).

By using these faces of oppression as filters through which to view classrooms I am working towards building a model of a mathematics curriculum for social justice in our schools. A project which is far from complete and often seems almost impossible to carry out.

I view such a project as worthwhile and even important. Social Justice offers a methodology which ties me to those values I held dear as a classroom teacher. I am not attempting to explain why classrooms work in the way they do - that seems like an even more impossible task - nor am I trying to offer a view of learning which can be generalised to improve our teaching.

"In short, what is sought is justice in education, based on a clear ideology that is neither oppressive, condescending nor patronising."

(Brandt, 1986, p. 122).

I have a great suspicion of people who tell me how children learn - it is always too easy to find a counter example. At the moment, this framework is offering me a way of asking questions which seem fundamental to me as a mathematics teacher and as a human being.

There ain't no justice, just us.

(Chumbawamba[2])

[2] Another popular British band of the 1990s

References

Bennison, A. et al.: 1984, 'Equity or equality: what shall it be?' in Fennema, E. and Jane Ayer, M. (eds), *Women and Education*, McCutchan Publishing Corporation, Berkely, pp 1-19.

Brandt, G. L.: 1986, *The Realisation of Anti-Racist Teaching*, The Falmer Press, Lewes.

Bourdieu, P. and Passeron, J.-C.: 1977, *Reproduction in Education, Society and Culture*, Sage Publications, London.

Collins, P. H.: 1990, *Black Feminist Thought*, Unwin Hyman, London.

Cotton, A.: 1994, 'Mathematics and a curriculum for justice' in *Proceedings of the BSRLM day conference, Saturday 3rd December, 1994*, London Institute of Education, pp 24 - 29.

Cotton, T. and Gates, P.: 1996, 'Why the psychological must consider the social in promoting equity and social justice in mathematics education' in *Proceedings of the 20th Conference of the International Group for the Psychology of Mathematics Education*, University of Valencia, Volume 2, pp 249 - 257.

Commission on Social Justice Report: 1994, *Social Justice, Strategies for National Renewal*, Vintage, London.

Daniels, N. (ed.): 1975, *Reading Rawls: Critical Studies of a Theory of Justice*, Basil Blackwell, Oxford.

Ernest, P.: 1989, *The Philosophy of Mathematics Education*, The Falmer Press, London.

Gilligan, C.: 1982, *In a Different Voice*, Harvard University Press, London.

Gilligan, C., Ward, J. V. and Taylor, J. M. (Eds): 1988, *Mapping the Moral Domain*, Harvard University Press, London.

Lucas, J. R.: 1980, *On Justice,* Oxford University Press, Oxford.

Lyotard, J. F.: 1992, *The Postmodern Explained to Children*, Turnaround, London.

McCarthy, C.: 1990, *Race and Curriculum: Social Inequality and the Theories and Politics of Difference in Contemporary Research on Schooling*, The Falmer Press, Basingstoke.

Okin, S. M.: 1991, 'John Rawls : justice as fairness - for whom?' in Shanley and Pateman (eds), *Feminist Interpretations and Political Theory*, Polity Press, Cambridge.

Pogge, T. W.: 1989, *Realising Rawls*, Cornell University Press.

Rawls, J.: 1971, *A Theory of Justice*, Oxford University Press, Oxford.

Ryan, A. (ed.): 1993, *Justice*, Oxford University Press, Oxford.

Restivo, S.: 1993, 'The social life of mathematics', in Restivo, S., Van Bendegem, J. P. and Fischer, R (eds), *Math Worlds - Philosophical and Social Studies of Mathematics and Mathematics Education*, State University of New York Press, Albany, pp 247 - 278.

Skovsmose, O.: 1994, *Towards a philosophy of Critical Mathematics Education*, Kluwer, Dordrecht.

Young, I. M.: 1990, *Justice and the Politics of Difference*, Princeton University Press, Chichester.

WHAT CAN SEMIOTICS DO FOR MATHEMATICS EDUCATION ?

Adam Vile

South Bank University

Abstract

There is developing interest in semiotics as a theoretical perspective in mathematics education. This paper examines semiotics from both a theoretical and practical perspective in order to begin discussion as to how and why semiotics may be a useful perspective for mathematics educators to adopt, or at least consider. A connection is drawn between the work of Vygotsky and Peirce and theoretical development provides the possibility of a consistent philosophical position that transcends Cartesian dualism and offers a new way of seeing. Then semiotics is examined practically from the point of view of a qualitative methodology. Finally suggestions are made about the role of theory in mathematics education and the possible role of semiotics as one of those theories.

Introduction

There is an inextricable connection between signs and mathematics. One might even say that mathematics consists entirely of a complex system of signs, a position that is defensible from within a semiotic perspective. Semiotics is the study of signs and sign functions in all conceivable aspects of message exchange; it concerns the conveyance and development of meaning through all sign vehicles. It is more a point of view than a method and is seen by some (Deely, 1990) in a more general sense, as a framework for the comprehension of the world. Interest in the nature of signs began with Aristotle, writers such as St Augustine and William of Ockham have had something to say on the subject but the term *semiotics* appeared first in 1690 in John Locke's *Essay Concerning Human Understanding*. Semiotics is a point of view on the world that puts the sign at the centre of all human action: a doctrine of communication. Intrinsic to the semiotic perspective is a dependence on the socio-cultural view of knowledge.

For some time now the mathematics education community has recognised the importance of the socio-cultural context of mathematics learning and such influences have become the object of study (Eccles and Jacobs, 1986; Walkerdine, 1990 and Evans and Tsatsaroni, 1994). Communication is the essential mediator of these socio-cultural influences and language is one (but not the only) sign system which effects this mediation. The role of language, both written and spoken, in mathematics has been investigated by Halliday (1978),

Klemme (1981) and Pimm (1987) amongst others, and it is evident from this work that language is instrumental in structuring and developing mathematical thoughts and actions. Nunes (1992) suggests further that specific cultural sign systems structure the organisation of mathematical activities but do not necessarily alter psychological functioning.

Currently there seems to be increasing interest in semiotic issues in mathematics education. A survey of the 1996 PME proceedings reveals one plenary (Puig 1996) and five research reports (Berenson and Vidakovic, 1996; MacNamara, 1996; Radford and Grenier, 1996; Redden, 1996; Vile and Lerman, 1996) in this area (this does not take into account the continuing interest in language and other more general socio-cultural issues which could certainly come under the remit of semiotics). Why is there such growing interest in semiotics? Is it just another bandwagon or does it (in some form or another) have the potential to add to the body of knowledge in mathematics education? In this paper I will consider these and other related questions taking as a context the domains of theory and practice.

Theoretical approaches

The increasing interest in cultural psychology is bringing to the fore the work of Vygotsky (1977) for whom semiotics was intrinsic to his description of the development of higher mental functioning. Vygotsky suggested first that the word and later that tools were the mediators from the intersubjective to the intrasubjective, and that concept development began in the social and was then internalised through the action of semiosis. Vygotsky and those working in his perspective put the focus of generalisation of higher mental functioning, specifically the generation of scientific concepts, on the sign. As a unit of psychological functioning he chose 'word meaning' and he gave to the sign the function of mediation, introducing a triadic scheme for meaning-making (Vygotsky, 1977).

Semiotics, however, did not begin with Vygotsky. It has two independent roots; one from structuralism and the work of Saussure, the other from the work of Charles Peirce the father of Pragmatism and of existential logic (Ayer, 1968). For Saussure signs signify by way of a signifier and a signified, i.e. a diadic action with direct connection. Peirce on the other hand introduced a triadic scheme (Hoopes, 1991) with a sign signifying an object only by way of a thirdness (related in some senses to Vygotsky's notions of mediation). In terms of both theory and analysis Saussurian semiologists prefer a synchronic approach, concentrating on the structure of differences in an act of communication rather than the signs themselves. Peircian semioticians propose a more dynamic perception of semiotic acts and analysis is more diachronic, paying proper attention to the history of the genesis of meanings. It would be fair to say that Peircian semiotics does not possess anywhere near the level of sophisticated tools as Saussurian semiology and as a result many of the same tools and much terminology finds its way into both paradigms. Choice between paradigms rests almost wholly on theoretical orientation and Peircian semioticians tend to be more philosophical than practical. Peircian semiotics is more appropriate for a cultural interpretation as it is diachronic in nature. Furthermore, it is an appropriate starting place for mathematics educators as it has many commonalities with the Vygotskian approach (Maffiolo, 1992, daws parallels between the two approaches). My own work adopts a Peircian perspective.

Peircian semiotics differs from that of Saussure in that Peirce sees the action between signifier and signified or (to use his terminology) sign and object, as triadic. The sign stands for the object in relation to a third element, the interpretant. It is the interpretant which introduces the cultural and historical dimensions to semiotics, but it is not at all easy to understand its role. Vygotsky too considered semiotic action to be triadic, yet for him the third element was the mediation means, the carrier of the meaning from external to internal. For Peirce the third element, the interpretant, although it may perform a mediating role is more than that. Peirce suggests that "[the interpretant] is all that is explicit in the sign itself apart from its content and circumstances of utterance" (Peirce, 1906, quoted in Deely, (1990, p. 26), and furthermore that the interpretant is a sign.

An example may serve to clarify the situation. Siegel and Carey (1989) describe a situation in which they show a preschooler a box of toothpaste and ask the child what the writing on the box says. The child replies "brush teeth". In this situation the sign (the box) stands for the object (toothpaste) in relation to the interpretant (the experience of having frequently brushed teeth after having seen the box). This is demonstrated diagrammatically in Figure 1. Many other replies were possible; for example the child could have associated toothpaste with going to the dentist or going to bed, or the sign could have indicated a number of different meanings to the child by providing for further context-dependent interpretants.

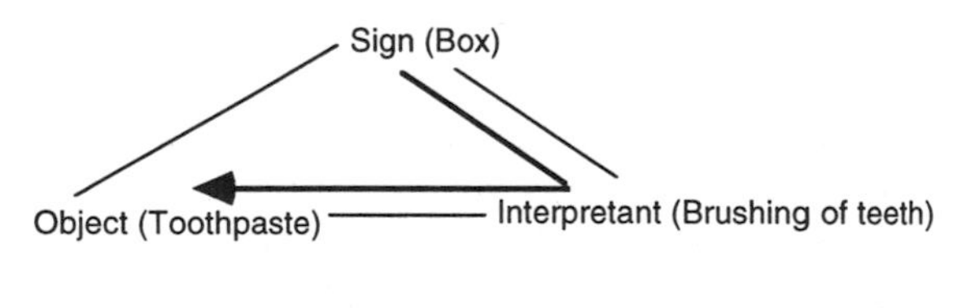

Fig 1

Interpretation of the sign may only take place by way of the interpretant - any direct connection between the sign and the object that is not recognised will remain devoid of 'sense'. The interpretant carries with it the socio-historical positioning of the interpreter and performs the functions of mediation and meaning-making. It is the single most important aspect of semiosis, for without it semiotic acts would have no meaning. The interpretant of a sign may become a sign in a new act of semiosis for which a further interpretant is needed in order to make sense of that act. In this way a process of "unlimited semiosis" (Eco, 1979, p. 69) begins evolving and developing the meanings of the sign diachronically, as a chain of signifiers (used by Walkerdine, 1988, in a very different sense to that in which it is used here. For Walkerdine this chain of signifiers represents a mainly synchronic shift in domains; here I use it in the sense of an historically developing system of dependent and compressed meanings from sign to interpretant to sign etc.) The very function of semiosis opens up a number of different meanings by providing for context dependent interpretants. Sign interpretation is bound up with everything we do, will do or have done and any given act of semiosis will be inserted into this network of experiences which accounts for the variation of individual interpretants. A sign is therefore not simply a substitute for a thing but contains some aspect of the meaning of that thing to us.

The introduction of the interpretant into the meaning-making equation also brings into question the nature of experience. Deely (1990) suggests that the semiotic universe divides the content of experience into three areas: the physical, the objective and the subjective. The objective is that which is experienced, the physical is that which is existent regardless of its being experienced and the subjective exists only in relation to the action of semiosis in which something that is objective is recognised and becomes a sign. The physical environment is thus experienced by the subject through a network of objective relationships which are then reconstituted by a subject in a subjective domain into an objectively shareable world. Through the interpretant the action of semiosis defines a three dimensional meaning-space which transcends subject-object dualism. Semiosis takes place primarily in a social space and interpretants may be thought of as “cultural units” (Eco, 1979) which both enable and are a product of a shared cultural understanding.

Vygotsky (1978) saw the development of higher mental functioning resulting from progressive control over semiotic systems of increasing demand; in this description thought is “nothing more than the self-appropriation of the cultural space” (Maffiolo, 1993). Vygotsky considered the word to be the cultural unit and saw the role of the word as central to cultural line development and specifically to concept development. Becker and Varelas (1989) have investigated the Vygotskian notion of semiotically mediated development. They conducted a study of counting and place value activities with increasing degrees of semiotic demand noting that as development occurs the sign-object connection gives way to a sign-sign connection, with signs becoming more opaque as development progresses.

In his description of the development of scientific concepts Vygotsky (1977) invokes a semiotic process by which "the relationship to an object is mediated from the start by some other concept. The very nature of a scientific concept implies its position in relation to other concepts" (Vygotsky, 1977, p.93). Scientific concepts are concepts that may be learned in a school context, things such as number and family relationships, and their nature is such that they depend upon other concepts (which from a Peircian point of view would be thought of as signs) for their meaning. This is consistent with the Peircian notion of unlimited semiosis (Eco, 1979) in which interpretants become signs which in turn become elements of further semiotic acts.

The works of Peirce and Vygotsky independently propose semiotics which transcends the subjective-objective divide through notions of thirdness. The interpretant provides the three-dimensional meaning-space where the intersubjective and intrasubjective meet to make meaning. Rather than consider the interpretant as the mediator from external to internal (as is the implication in the writing of Vygotsky, 1977; Eco, 1979 and Ohtani, 1994) the role of the interpretant is extended to be a space in which meanings are constituted and re-constituted both publicly and privately.

> With Peirce and Vygotsky, [we] recognise the subject as a sign and the sign as a cultural object, we place ourselves in a dialogical perspective, where thought is elaborated by and in the activity of communication, where cognition is constructed by and in the share of speech codes and contexts. This is in fact the opposite of traditional epistemology, which assumes that human beings use signs to communicate the thoughts they previously and individually constructed according to natural and/or universal processes. (Maffiolo, 1991, p. 491).

My own work has been concerned with the evolution of a theoretical framework which synthesises diachronic Peircian semiotics with the developmental aspects of the work of Vygotsky into a coherent theoretical perspective with sound philosophical underpinnings. Developmental semiotics (described in more detail elsewhere: Vile, 1996b; Vile and Lerman, 1996) has evolved to incorporate re-defined and refined notions from the work of Peirce and Vygotsky. Central to this perspective are the assumptions that mathematical meaning-making is social in nature and that any account of mathematical meaning-making should take account of development.

There are two main elements of the process of developmental semiotics. The first, from Peirce and Deely, describes the way that a cognising subject is in possession of a network of experiences into which any semiotic act will be inserted and an interpretant (and hence a meaning) will be made. This network of experiences is built up over time with each new experience affecting the whole network. The second, from Vygotsky (1977), puts forward the idea of a developmental process in which the external social factors affect the level of mental functioning and force the transition from elementary to higher mental functions. Meanings are made through the sign-object-interpretant action, the interpretant being the element of thirdness that transports a physical, sign-object brute act of secondness into the objective, social and, subsequently, intrapersonal plane. The interpretant is the indicator of the meaning. Each interpretant itself becomes a sign and may enter, at the level of secondness, into a further act of semiosis. In this way an unlimited semiosis tending towards sign-sign (symbolic) functioning will be built into the network of experiences. The overall result will be a developmental shift towards the more abstract with the object referent moving more to the background and sign-sign functioning becoming foregrounded (Becker and Varelas, 1989).

As well as providing a theoretical perspective, a lens through which to view mathematical meaning-making, developmental semiotics provides a vocabulary for theoretical and practical description and discussion of actual and virtual meaning- making processes. Important in practice are notions of *semiotic demand* (relating to the level of opacity of a given sign) *sign-sign foregrounding* (relating to the degree of generalisation of a sign held by an individual and measured by their ability to access a sign with a given semiotic demand), and (sign) *meaning* (or interpretant). I would like to suggest that developmental semiotics may be fruitful both in descriptions of meaning-making and as a tool for qualitative description of the meaning-making process.

Semiotics as a qualitative methodology.

Semiotics as a philosophical perspective is concerned with the world as a web of signs (Deely, 1990) and semiotic analysis is concerned with the reading of those signs. In this respect research within the qualitative paradigm (e.g. symbolic interactionism, phenomenology, ethnography) with its search for meaning in action and culture could be considered to fall within the semiotic perspective. Shank (1995) suggests that:

> Semiotic theory can help expand the conceptual and practical domain of qualitative research by serving as a philosophical foundation of the discipline, thereby allowing qualitative researchers to build upon a set of ideas that powerfully extends the aims and goals of their research. Qualitative research in education can help expand semiotics by serving as a source of empirical research and findings, thereby helping move semiotics away from its current near total preoccupation with theory and into a state where empirically determined issues play a more important and visible role. (para. 2).

Semiotics does possess a number of tools for analysis of qualitative data but as Shank says the emphasis has been more on theory than practice.

Semiotic chains, semiotic clustering and semiotic squares are amongst the most common techniques used in semiotic analysis. Manning (1987) and Feldman

(1995) outline these and other tools arising from a structuralist semiotic perspective. An excellent article for beginning semioticians by Chandler (1996) identifies a number of aspects of semiotic analysis drawn from both diachronic and synchronic points of view. In a section entitled "D.I.Y" semiotic analysis he suggests a number of guidelines for the analysis of media and text which I have adapted somewhat here for the sake of generality (Box 1). Text here relates not only to written text but any system of semiotic action, a text could be a classroom dialogue, a piece of writing or a physical action.

As an example of the use of these criteria consider an analysis of the following "text" (Fig 2) which was constructed by a student, who shall be known as Becky. The data were collected during a case study in the area of algebra.

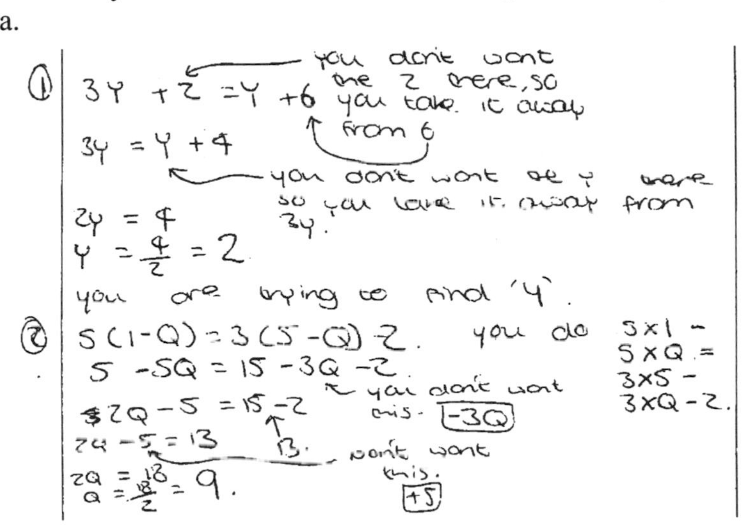

Fig 2: Becky's "text"

1 **What are the important signifiers and what do they signify?**

- What is the system within which these signs make sense?
- What connotations seem to be involved?

2 **What is the syntagmatic structure of the text?**

- How does one unit (action, part of the text) relate to another?
- Are there formulaic features that have shaped the text?

3 **Paradigmatic analysis**

- To which class of paradigms (medium; genre; theme) does the whole text belong?
- What might the text have been like if it had formed part of a different genre?
- What paradigms are noticeably absent?
- What paired opposites seem to be involved (e.g. nature/culture)?
- Is there a central opposition in the text?
- What psychological, social and political import do these oppositions have?

4 **What semiotic codes are used?**

- Which conventions are most obvious in the text?
- Which codes are specific to the genre of text (writing, acting, speaking)?
- Which codes are shared with other genre?
- What cultural assumptions are called upon?
- What seems to be the preferred reading?
- How far does this reflect or depart from dominant cultural values?
- What alternative readings seem possible?

5 **Intertextuality**

- Does it allude to other genres?
- Does it allude to or compare with other texts within the genre?
- How does it compare with treatments of similar themes within other genres?
- What other contributions have semioticians made that can be applied to the text?
- What does a purely structural analysis of the text downplay or ignore?

Box 1: Guidelines for the analysis of media and text

1 What are the important signifiers and what do they signify?

This text is replete with signifiers, some from the system of mathematics and some from the system of the English language. There are two different sign types, *symbols* (e.g. X, 2, don't) and *indexes* (the arrow) and it is interesting to note that the index is used in this case only when reflecting or describing, not when doing.

2 What is the syntagmatic structure of the text?

Algebra by its very nature has an intrinsic "grammar" and structure. Becky's work is structured along these lines - work is sequential and reflections describe the processes passed through from step to step. Becky is telling the story of how she solved the equation and in a syntagmatic sense this is the same story for both equations: equation is written; equation is expanded to a recognisable form; equation is manipulated to the form $ax = b$ by way of addition and subtraction of appropriate terms; unknown is stated. This is perhaps a basic protocol or genre for the solution process that Becky is using.

3 Paradigmatic analysis

This text perhaps belongs to a syntactic solution paradigm, but not entirely as there is a sense in which her manipulations could be endowed with more meaning, presenting a syntactic/semantic complementarity. Paradigmatic analysis usually proceeds by identifying oppositions in the text. I have identified a complementarity yet an opposition does exist in this text between the knower and the known (or in this case the unknown). Becky knows that she is "trying to find *y*", but she does not know what it is, although *y* exists as a known by virtue of the structure of the equation. For Becky it seems that in this instance the value of *y* is to be uncovered; in a sense it is known and in a sense it is not.

4 What semiotic codes are used?

There are codes specific both to mathematics and to the medium. In a sense this sort of mathematics would be very difficult to do without visual coding, but by the same token a computer screen could generate a different use or meaning of the codes used. Clearly Becky is using signs that are culturally accepted as appropriate, but it is not clear whether she shares the same code as others. All we have to go on is the fact that the protocol is followed accurately. One reading of the text is that Becky is acting meaningfully in a context in which she feels comfortable. Another is that she is senselessly rule-following.

5 Intertextuality

This analysis can only really be achieved through comparison with other extracts from other students solving the same equation. In addition there are elements of the text relating to the diachronic nature of Becky's learning which have been ignored. A full intertextual analysis is only possible if history is included in the equation.

Overall this analysis is structural and synchronic, yet it has brought to light a number of different elements of the text which perhaps were not explicit. However this type of structuralist analysis has been criticised (Deely, 1990) because of its synchronic approach and lack of consideration of cultural and societal issues. It is still valuable (but perhaps not as revealing) to construct the analysis of a text in isolation. An alternative analysis must consider the text as one part of a history of connected texts and experiences. A consistent set of tools for this type of analysis does not yet exist but one may begin to interpret the text from within a developmental semiotic point of view.

How might one begin to analyse the extract shown above from the point of view of a developmental semiotic? One of the basic tenets of developmental semiotics is that extracts like this should not be taken in isolation and so to give a brief history of this student's algebraic action would be appropriate and perhaps revealing. Becky had a very successful beginning to her experiences with linear equations. Equations such as $x + 5 = 7$ were solved with recourse to trial and error methods like the ones shown in the following extract (which does not come from the pen of Becky, but illustrates the general principle and will be used as a comparative example later on in this discussion):

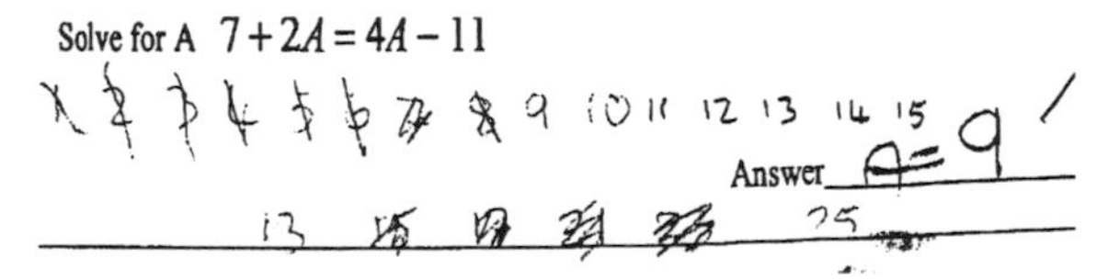

Fig 3: Trial and error method

Equations such as the one above (Fig 3) were not successfully solved by Becky as she was forced to progress from trial and error methods by the increased semiotic demand of the equations into an attempt to manipulate symbols to find the value of the unknown. Equations were increased in semiotic demand until we reach the stage about which Becky is writing in the extract given above. By this stage she has become more successful. She explains how to solve an equation with the unknown on both sides. This is an equation with a high degree of semiotic demand. In other words it is an equation in which the signs do not stand directly or transparently for existent objects (rather they represent a chain of signification from protonumbers to arithmetic) and as a result potentially may be thought of as objects in their own right. Becky states "you don't want the *Y* there" (I was not able to ask why not although I would have liked to) and demonstrates her ability to "take *Y* away from 2*Y*", in order to isolate the unknown, that is to treat the sign as an object in its own right (Pimm, 1995). The referent of the sign (whatever that is to her) has been backgrounded and the connections with other signs have been foregrounded, giving rise to a degree of sign-sign foregrounding that enables Becky to manipulate, solve and reflect upon the equation.

The second equation is more problematic. There is a higher degree of semiotic demand since this equation consists of a complex string of symbols with associated rules of manipulation. The problem occurs in the third line where Becky misses out a number of steps and writes down an identity which does not follow from the previous line. She produces the answer 9 but does not substitute to check that this satisfies the equation. In treating the sign as an object, Becky has become over-confident and as a result has made an error in "mental algebra". The real problem is that she seems not to see the connection between the symbol for the unknown and the number which is the answer to the equation. In a pre-study interview when asked what the x stands for in the equation $x + 2 = 5$ she said, "x is something you don't know what it is".

Compare this with the example above in which A is established as 9 by a trial and error method. This student does not possess the required degree of sign-sign foregrounding to consider the unknown as an object and to manipulate it, but she does possess a well developed meaning of the nature of the unknown - as a number. The equation has a high degree of semiotic demand yet this student is able to solve it by virtue of a more transparent (less developed?) notion of the sign.

This type of analysis can be applied to individual cases and to larger case studies. It is not at all difficult to see how such analysis, with its search for meaning and focus on the sign, has the potential to inform future practice. In the case of Becky steps must be taken to introduce the notion that X stands for a general number, and in the case of the second student steps must be taken to force the construction of X as an object in its own right, perhaps through the introduction of equations with non-integral solutions.

It is not the answering but rather the asking of questions from within a semiotic perspective which may be revealing, for the focus of these questions is the reading of a text and of signs within that text. Educational researchers using semiotic analysis need to become good readers of signs, but that is not to say that the readings or interpretations that they make describe with any degree of "trueness" the underlying structures, meanings or connotations manifested in those signs. Questions may be raised about the value of semiotic analysis if one accepts the relativity and subjectivity of the techniques. But this is not problematic if one sees semiotics not as a methodology for extracting descriptions of the way things are but rather as a way of "illuminating" (Feldman, 1995, p. 39) relationships and structures, that is as a toolkit to enable the creation and evolution of interpretations that make new sense of data in a way that develops our own understandings of phenomena and requires us to ask and attempt to answer new questions. In providing a set of empirical tools for the analysis of data semiotics presupposes theoretical tools of equal importance which carry with them a register, a vocabulary, a way of seeing and talking that will enable discourse and the creation of interpretations.

Qualitative research from a semiotic point of view is not aimed at extracting facts but at describing the world through the action and interpretation of signs. I like to use the metaphor of semiotics as a lens through which to view and interpret the world, and a scheme for research based upon this world view may provide a systematic, empirical methodology with a strong theoretical underpinning that may prove fruitful and revealing in all areas of social research. I have shown one example of how this type of analysis may be applied in research but one may ask whether there is indeed any use of semiotics in the classroom.

Teachers have so much to cope with nowadays, so many new curriculum documents and ideas, that they hardly have time to think about anything else apart from teaching and administration. It has been suggested (Scott-Hodgetts, 1988) that teachers should be interested in research in mathematics education and that indeed they should actively be engaging in research in their own classrooms. Lerman (1990) introduces the notion of the reflective practitioner to the classroom and in this sense it is clear that every teacher engages in research of some sort, perhaps in testing new materials or teaching ideas or perhaps in interpreting pupils' actions. Scott-Hodgetts (1990) suggests that, for the teacher, involvement in research helps develop a more acute awareness of the activities around teaching and learning in the classroom. It is in this that I believe semiotics has a part to play.

As an interpretative framework with focus on meaning rather than knowledge and with a body of pre-existing theoretical and practical ideas, semiotics has the potential to provide insights into the ways in which students make meaning in the mathematics classroom. What semiotics does is change the focus of interpretation and inquiry to the sign. In the most simple case teachers can ask of themselves: what does that sign mean to that student in that setting? Questioning, observation and experience may give rise to answers which would lead to solutions, to ways of helping students redefine for themselves the nature of the signs that they are using in a way that will help them cope with signs of higher semiotic demand. Additionally a semiotic perspective raises questions about the nature of teaching and learning. With focus on the sign, teaching is viewed as a process of facilitating meaning-making rather than imparting knowledge and this in turn brings to the fore notions about the relative nature of our subject. Semiotics may offer teachers both a valuable interpretative framework and a challenge.

Of course if one subscribes to this view one must accept that mathematics consists entirely of signs, that is it is a semiotic system (Rotman, 1995) and that, for what ever reason, it progresses by producing ever more abstract signs with ever more increasing semiotic demand and that a teacher's role is in part to give access to those signs and that semiotic system. If these premises are accepted then it becomes imperative for teachers to begin some sort of semiotic analysis on the mathematics to which they intend to expose their students and of their individual reactions and actions and hence meanings. There is no need (for example) to ask questions about the syntagmatic and paradigmatic aspects of the text, but such terminology may prove useful when experiences and interpretations are shared with others. Semiotic analysis occurs in the classroom today, it is just not labelled as such, and a more systematic approach, such as that presented in this paper may be more helpful. The aim always is to read the signs and then, if appropriate and efficient, effect change.

Semiotic analysis is not confined to the mathematics aspect of the mathematics classroom. One of the key elements in semiotics is the experiences that are brought to each semiotic act (Deely, 1990). Students bring experiences from home, from the playground, from television etc. into the classroom and these experiences will have an effect on the way in which they interpret and act with the signs all around them (even the mathematical ones, as Walkerdine, 1989, has shown). Semiotic analysis of small classroom incidents could give clues to a larger socio-cultural picture.

So What Can Semiotics Offer Mathematics Education?

Before this question can be fully answered the role of theory in mathematics education must be considered. I suggest that theory is not only useful but essential in our attempts to make sense of the world in which we exist, and of the observations that we make. Feyerabend (1975) contends convincingly that observational statements presuppose theory, that although natural observations may be made, attempts to make sense of them, describe them and share them require some kind of theory. Clearly there are strong grounds for the acceptance of this view in both science and social science, and the implication is that in order to observe and interpret classroom phenomena some theory is necessary. Additionally, the sharing of a common theoretical perspective would perhaps provide a context and vocabulary for discussion and interpretation within a community. Mathematics education has within its bounds a number of such theories and mathematics educators have the possibility and facility to pick and choose, and use these theories in acts of observation, interpretation and prediction. The development of such theories aims to improve their ability to provide fruitful observation statements and to make new and useful interpretations and this development is an important aspect of mathematics education research in itself contributing to the evolving dialectic of theory and practice. Semiotics is such a theory, in mathematics education terms it is only in its embryonic stages but already it is (as I hope that I have shown here and elsewhere) able to contribute both theoretically and practically to the general body on mathematics education research and practice.

I would like to suggest that semiotics may be of use for mathematics educators (and also for other researchers in qualitative paradigms) in both a reconceptualisation of theoretical perspectives and in terms of empirical methodologies. Theoretically, semiotics may offer a description of the meaning-making process that accounts for the socio-cultural nature of experience and, avoiding Cartesian dualism and with it problems with intersubjectivity (Lerman, 1996), provides an alternative description of the social/personal interface. Empirically, focus on the role of the sign, and in particular the nature of the interpretant, may assist in the understanding of the meaning-making process in specific contexts. Continuing work would necessarily build on work carried out in specific semiotic systems such as that of Klemme (1981), Pimm (1987), Rotman (1988), Stage (1991) and Nunes (1992), and may serve to unite such work under a comprehensive semiotic framework. I hope that the discussion of the theoretical aspects of a semiotic point of view will be sufficient to generate interest from the mathematics education community in the possible application of a semiotic framework to the description of meaning-making in mathematics.

In the current climate, mathematics educators are moving towards a description of mathematical meaning-making that involves society, culture, communication and context, and many are moving further to investigate the role of the sign in mathematical meaning-making. Semiotics, at least in versions presented by Peirce and Vygotsky, exists as a well developed point of view of the world that embraces all of these elements in a way that transcends traditional philosophical objections and as such may prove (and indeed is already proving) to be a fruitful perspective for the description of meaning-making in mathematics education in particular.

References.

Ayer, A.: 1968, *The Origins of Pragmatism,* Macmillan.

Becker, J. and Varelas, M.: 1989, “Semiotic aspects of cognitive development: illustrations from early mathematical cognition’, *Psychological Review* **100**(3), 420-431.

Berenson, S. B. and Vidakovic, D.: 1996, “Children's word meanings and the development of division concepts’, in *Proceedings of the 20th Conference of the International Group for the Psychology of Mathematics Education*, University of Valencia, Volume 2, pp. 75-80.

Chandler, D.: 1996, *Semiotics for Beginners,*: *http://www.pccp.com.ar/semiotic.html*

Deely, J.: 1990, *The Basics of Semiotics*, Indiana University Press, Bloomington.

Deely, J.: 1993, ‘Reading the signs: some basics of semiotics’, *Semiotica* **97**(3), 247-266.

Eccles, J. and Jacobs, J.: 1986, ‘Social forces shape maths attitudes and performance’, *Signs: Journal of Women in Culture and Society*, **11**(2), 367-380.

Eco, U. 1979, *A Theory of Semiotics,* Indiana University Press.

Evans, J and Tsatsaroni, A.: 1994, "Language and subjectivity in the mathematics classroom', in Lerman. S. (ed.), *Cultural Perspectives in the Mathematics Classroom*, Kluwer, pp. 169-190.

Feldman, M.: 1995, *Strategies for Interpreting Qualitative Data,* Sage.

Feyerabend. P.: 1975, *Against Method*, Verso.

Halliday, M. A. K.: 1978, 'Sociolinguistic aspects of mathematical education: language as social semiotic' in Halliday, M. A. K., *The social interpretation of language and Meaning*, Arnold, pp. 194-204.

Hoopes, J.: 1991, *Peirce on Signs, Writings on Semiotics by C. S. Peirce*, University of North Carolina Press.

Klemme, S. L.: 1981, 'References of speech acts as characteristics of mathematical classroom conversation', *Educational Studies in Mathematics* **12**, 43-58.

Lerman, S.: 1990, 'The role of research in the practice of mathematics education', *For the Learning of Mathematics* **10**(2), 25-28.

Lerman, S.: 1996, 'Intersubjectivity in mathematics learning: a challenge to the radical constructivist paradigm', *Journal for Research in Mathematics Education* **27**(2), 133-150.

MacNamara, O.: 1996, 'Mathematics and the sign', in *Proceedings of the 20th Conference of the International Group for the Psychology of Mathematics Education*, University of Valencia, Volume 3, pp. 369-378.

Maffiolo, D.: 1993, 'From a social to a cultural approach in the study of cognitive activities: the fundamental role of semiotic systems', *European Journal of Psychology of Education* **8**(4), 487-500.

Manning, P.: 1987, *Semiotics and Fieldwork,* Sage.

Nunes, T.: 1992, 'Cognitive invariants and cultural variation in mathematics concepts', *International Journal of Behavioural Development* **15**(4), 433-453.

Pimm, D.: 1987, *Speaking Mathematically*, Routledge, London.

Pimm, D.: 1995, *Symbols and Meanings in School Mathematics,* Routledge, London.

Puig, L.: 1996, 'Pupils' prompted production of a medieval sign system', *Proceedings of the 20th Conference of the International Group for the Psychology of Mathematics Education*, University of Valencia, Volume 1, pp. 77-87.

Radford, L. and Grenier, M.: 1996, 'On dialectical relationships between signs and algebraic ideas', *Proceedings of the 20th Conference of the International Group for the Psychology of Mathematics Education*, University of Valencia, Volume 4, pp. 179-186.

Redden, E.: 1996, 'Wouldn't it be good if we had a symbol to stand for a number?', *Proceedings of the 20th Conference of the International Group for the Psychology of Mathematics Education*, University of Valencia, Volume 4, pp. 195-202.

Rotman, B.: 1988, 'Towards a semiotics of mathematics', *Semiotica,* 72-1/2, 1-35.

Rotman, B.: 1995, *Ad Infinitum: the ghost in Turing's Machine*, Stanford University Press.

Scott-Hodgetts, R.: 1988, 'Why should teachers be interested in research in mathematics education?', in Pimm, D. (ed.), *Mathematics, Teachers and Children,* Hodder and Stoughton.

Shank, G.: 1987, 'Abductive strategies in educational research', *The American Journal of Semiotics,* special issue on Semiotics and Education **5** 175-190.

Shank, G.: 1995, 'Semiotics and qualitative research in education: the third crossroad', *The Qualitative Report*, **2**(3), *:http://www.nova.edu/ssss/QR/*

Siegel, M. and Carey, R.: 1989, *Critical Thinking: A Semiotic Perspective*, Monographs on Teaching Critical Thinking, Volume 1, National Council of Teachers of English, Bloomington.

Stage, F.K.: 1991, 'Semiotics in the college mathematics classroom', *Paper presented at the annual meeting of the American Educational Research Association*, Chicago.

Vile, A.: 1996a, "Peirce, the interpretant (a tripartite division of experience) and mathematical meaning', *Proceedings of Working Group 10, ICME 8,* Seville, Spain.

Vile, A.: 1996b, *Developmental Semiotics: The evolution of a theoretical perspective for the description of meaning-making in mathematics education and mathematics,* Unpublished PhD dissertation, South Bank University, London.

Vile, A and Lerman, S.: 1996, "Semiotics as a descriptive framework in mathematical domains', *Proceedings of the 20th Conference of the International Group for the Psychology of Mathematics Education,* Volume 4, pp. 359-402.

Vygotsky, L.: 1977, *Thought and Language*, M.I.T. Press, Cambridge MA.

Vygotsky, L.: 1978, *Mind in Society,* Harvard University Press, Cambridge MA.

Walkerdine, V.: 1989, *The Mastery of Reason: Cognitive Development in the Production Of Reality*, Routledge, London.

EXAMPLES, GENERALISATION AND PROOF

Liz Bills* and Tim Rowland**

University of Warwick* and Institute of Education, University of London**

Abstract

The interplay between generalisations and particular instances - examples - is an essential feature of mathematics teaching and learning. In this paper, we bring together our experiences of personal and classroom mathematics activity, and demonstrate that examples do not always fulfil their intended purpose (to point to generalisations). A distinction is drawn between 'empirical' and 'structural' generalisation, and the role of generic examples is discussed as a means of supporting the second of these qualities of generalisation.

In[tro]duction

For all learners of mathematics there is the possibility of acquiring new knowledge by reflection on appropriate and relevant experience (and arguably there is no other way). Unifying and information-extending insight is central to such a means of coming-to-know, and may be viewed as a form of generalisation. The type of generalisation with which we are particularly concerned here may be described as 'generalisation from cases'. MacLane (1986) uses this term to mean a generalisation which subsumes several particular cases e.g. the expansion of $(a+b)^n$ as a generalisation of $(a+b)^2$, $(a+b)^3$ etc. In this example the general rule summarises some features of the specific cases. It also claims the plausibility of the generalisation to cases beyond those which have been examined, bridging the 'epistemic gap' between the known and the unknown. This feature of the general rule we describe as 'inductive reasoning'. For the great mathematicians, as well as for novices, mathematics characteristically comes into being by inductive intuition, not by deduction.

> Analysis and natural philosophy owe their most important discoveries to this fruitful means, which is called induction. Newton was indebted to it for his theorems of the binomial and the principle of universal gravity. (Laplace, 1902, p. 176)
>
> I must admit that I am not in a position to give it a rigorous demonstration [...] The examples I have just developed will undoubtedly dispel any qualms which we might have had about the truth of my formula. (Euler, translated by Polya, 1954, pp. 93-95)

> The purpose of rigour is to legitimate the conquests of the intuition. (Hadamard, quoted by Burn, 1982, p. 1)

The products of induction are plausible 'truth-estimates' (Rescher, 1980, p. 9), and such conjectures may well be held with conviction. But whereas initial regularity is so often a reliable guide to generality in mathematics, it is not invariably so. Consider the (false) propositions that $n^2 + n + 41$ is prime for all n (true for $n = 1$ to 39), and that the number of regions of a circle formed by joining each of n points (irregularly spaced) on the boundary to every other is a power of 2 (true for $n = 1$ to 5). In these two examples, the mere accumulation of confirming instances misleads. We shall argue that the quality of such evidence is weak for mathematical generalisation, and indicate the need for other sources of conviction.

Conviction and scepticism

Stamp (undated) recalls teaching a lesson on right-angled triangles. In the first two examples considered - (6, 8, 10) and (5, 12, 13) - it was observed that the area and perimeter had the same numerical value. This led to the conjecture that "this happens every time". Stamp reports that he "denied" that this can be so, and in fact proceeds in the note to deductive demonstration that, with the exception of the given examples, the proposition is universally false.

The mathematics teacher's reaction to the conjecture may well be that if there were such a connection between the perimeter and area of integer-sided right-angled triangles, then they would already know about it! Therefore (they might reason) there can be no such connection. Whilst this 'mature' mode of reasoning *can* be a reliable guide to induction, it can also be a negative and dangerous reason for scepticism about the remarkable-but-unfamiliar.

Induction is essential for mathematics, but it is not sufficient; the "conquests of the intuition" are potentially fallible.

We report two incidents recorded by a teacher of lower sixth A-level mathematics students.

> During a reporting-back session following an exploration of the absolute value function, Lorne makes the assertion that the graph of $y = |f(x)|$ is the same as the graph of $y = f(|x|)$ for every function f. I am unsure whether he is right and I try to think of counter-examples. I suggest that he plots $y = |2x + 1|$ and $y = 2|x| + 1$.

Lorne had considered five or six examples of functions in coming to this inductive conclusion. The teacher was doubtful, but not because the number of examples considered was too small. She felt that such a striking result would already have been known to her. She also had an image of the graphs of modulus functions that involved points with undefined gradient, where the graph was "reflected back on itself". The initial, vague feeling of unease formed itself into a counter-example. Then she could see that Lorne's examples may have been sufficient in number but of "the wrong kind".

Whatever her intuition was which made her doubt the truth of Lorne's statement, there was no such doubt in Lorne's mind. We account for this difference in two ways. First, he had less experience of the modulus function on which to draw. Secondly, he was less cautious of inductive reasoning. His schooling had often put him in the position of needing to trust conclusions from inductive reasoning in mathematics without considering the strength of other reasons for conviction.

A few days later the teacher recorded what she saw as a similar incident from a class lesson:

> I am talking to the whole class about the way in which they derived the equation of a circle with radius 2 and centre (3, 5). I have written the equations $\sqrt{(x-3)^2+(y-5)^2} = 2$ and $(x-3)^2 + (y-5)^2 = 4$ on the board. I ask "where did the 3 the 4 and the 5 come from in this (the second) equation?" Trevor replies that the 4 is the diameter of the circle.

The teacher had intended to draw the students' attention to the *structure* of the derivation of this particular equation, with the eventual aim that they would appreciate the form $(x-a)^2 + (y-b)^2 = r^2$ for the equation of a circle. She was expecting them to base an answer to her question on their recall of the *procedure* by which they had derived the equation. But Trevor seemed to be making an *empirical* generalisation from one case, rather than recalling the derivation procedure as the teacher had hoped. We see his statement as a generalisation because he said that 4 was 'the diameter of the circle' and not simply twice 2. His answer relied on seeing that 4 was twice the radius, rather than seeing that 4 was the radius squared, and that it resulted from the squaring operation which was part of the process of obtaining the equation. He focused his attention on *numerical patterns* rather than *structural relationships*. By contrast, the teacher knew that the constant term in this equation could not, in general, be equal to the diameter of the circle, since she knew it to be equal to the square of the radius.

Empirical and structural generalisation

In both of the cases above, the student made a (sometimes tentative) inductive generalisation. In each case, the teacher was sceptical about their conclusions on the grounds of other sources of conviction. In the second, Trevor generalised from the particular circle equation in a way which was different from that intended. We use the terms 'empirical' and 'structural' to describe, respectively, the form of generalisation Lorne and Trevor made and that which the teacher had expected. In using these terms, we emphasise that one form of generalisation is achieved by considering the form of results, whist the other is made by looking at the underlying meanings, structures or procedures.

This distinction is illuminated by Liz's notes on the following problem:

> **Problem** The picture (right) shows a rectangle made up of two rows of four columns and of squares outlined by matches. How many matches would be needed to make a rectangle with R rows and C columns?

When I first worked on this problem, I decided to simplify by holding the number of rows constant. I held R as 2 and produced a series of diagrams such as these:

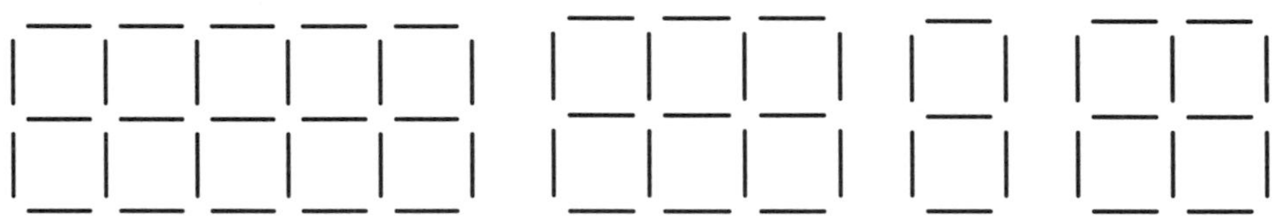

From my diagrams I produced the following table of results:

No. of columns (C)	No. of matches (M)
4	22
5	27
3	17
1	7
2	12

I saw that the results in my table fitted the rule $M = 5C + 2$. My trust in this formula for all positive whole number values of C was based first on the results in my table. Secondly, I was confident in it because it was of the form I was expecting. By this I mean that I expected a relationship to exist between M and C, and previous experience led me to expect the relationship to be linear.

Next I changed the value of R to 3 and, with the aid of one or two diagrams, convinced myself that M and C now fitted the rule $M = 7C + 3$. Similarly, I found that, for $R = 4$, $M = 9C + 4$ and, for $R = 5$, $M = 11C + 5$. For these, I needed fewer diagrams and tabulated results, because my conviction about these formulae from sources other than my table of results was greater each time. Having established that linear relationships held for $R = 2$ and $R = 3$, I needed only two results in the case $R = 4$ in order to be convinced that I had the correct formula.

Now a pattern was emerging that suggested that a general rule was $M = (2R + 1)C + R$.

Again, in moving from these separate formulae for different values of R to one which incorporated variations in R, I based my conviction first on the four formulae I had identified in the special cases $R = 2, 3, 4, 5$. But I had also the anticipation that such a general formula would exist, would be linear in R and in C and would be symmetrical with respect to the two variables.

Finally, seeing beyond the particular numerical features of a diagram so as to perceive it as a 'generic' representative of the general (see below), I was able to see that I could count the number of vertical and horizontal matches as follows:

> there are $C + 1$ columns of vertical matches, each containing R matches;
> there are $R + 1$ rows of horizontal matches, each containing C matches;
> therefore there are altogether $(C + 1)R + (R + 1)C$ matches.

This line of argument confirmed the rule which had been arrived at empirically. The argument is *structural*, because it is based on a way of counting the matches in this configuration. By contrast, the first line of argument is *empirical* because it is based (predominantly) on a pattern in the table of results: it argues 'for small values of R and C the number of matches is given by $(C + 1)R + (R + 1)C$ so it seems reasonable that this will be the case for all positive whole number values'. Inductive reasoning of this first kind can be a useful way of *producing* conjectures, but in the absence of other sources of conviction it may point to erroneous conclusions.

Generic examples

The story (probably apocryphal, but see Polya, 1962, pp. 60-62 for one version) is told about the child C. F. Gauss, who astounded his village schoolmaster by his rapid calculation of the sum of the integers from 1 to 100. Whilst the other pupils performed laborious addition, Gauss added 1 to 100, 2 to 99, 3 to 98, and so on, and finally computed fifty 101s with ease. The power of the story is that it offers the listener a means to add, say, the integers from 1 to 200.

Gauss's method demonstrates, by 'generic example', that the sum of the first $2k$ positive integers is $k(2k + 1)$. Nobody who could follow Gauss' method in the case $k = 50$ could possibly doubt the general case. It is important to emphasise that it is not simply the *fact* that the proposition that the sum $1 + 2 + 3 + ... + 2k = k(2k + 1)$ has been verified as true in the case $k = 50$. It is the *manner* in which it is verified, the form of presentation of the confirmation. As Balacheff (1988) so clearly and elegantly puts it:

> The generic example involves making explicit the reasons for the truth of an assertion by means of operations or transformations on an object that is not there in its own right, but as a characteristic representative of the class. (p. 219)

Sometimes, structural generalisation can be achieved by a generic example. This mode of proof is also discussed by Mason and Pimm (1984):

> The generic proof, although given in terms of a particular number, nowhere relies on any specific properties of that number. (1984, p. 284)

Closely related to proof by generic example, is the notion of 'action proof' (Semadeni, 1984; Walther, 1984). The generic example serves not only to present a confirming instance of a proposition - which it certainly is - but to provide insight as to why the proposition holds true for that single instance. For an action proof to be effective it is necessary for the student (or other audience) to identify those aspects of special examples which are "invariant regarding a transfer to other arbitrary examples" (*op. cit.* p. 10). The transparent presentation of the example is such that analogy with other instances is readily achieved, and their truth is thereby made manifest. The intention is that the audience can ultimately conceive of no possible instance in which the analogy could not be achieved. Despite the designer's generic pedagogic intention, such an argument may not necessarily be received by the student with the intended generality. Trevor's perception of the equation of the circle with radius 2 and centre (3, 5) is a case in point.

Our evidence in this respect is mixed: some first-year undergraduate Mathematics/Education students, following a 16-hour module on process aspects of mathematics, were introduced to the well-known 'Stairs' investigation. This problem concerns the number of ways of ascending a flight of n stairs in combinations of ones and twos. The Fibonacci sequence readily emerges in the data, and these students were asked to consider why this is the case - in effect, whether the obvious inductive inference is as valid as it is convincing. One student, Kim, gave an account of why it is that the number of ways for 6 stairs will be equal to the sum of the number of ways for the previous two numbers of stairs. To investigate whether Kim's explanation was perceived as particular or generic, the students were asked to complete the questionnaire below (with spaces for students to write their responses):

Climbing stairs in ones and twos

Observation: The number of ways for 6 stairs [**13**] is equal to the sum of the number of ways for 5 stairs [**8**] and the number of ways for 4 stairs [**5**].

Explanation: This is because, in ascending 6 stairs, the first step must be a one or a two. If it is a one, there are 5 stairs left, and there are **8** ways of climbing these 5 stairs. If it is a two, there are 4 stairs left, and there are **5** ways of climbing these 4 stairs. Therefore there are **8 + 5** ways of ascending 6 stairs.

1 Are you happy with the above explanation i.e. is it convincing?

2 Does the above explanation help to convince you that the number of ways for 15 stairs will be equal to [the number of ways for 14 stairs] + [the number of ways for 13 stairs]?

If you answered YES, why does the explanation for 6 stairs convince you for 15 stairs?

If you answered NO, what would you need, in order to be convinced?

3 Does the first explanation [for 6 stairs] convince you that the number of ways *for any number* of stairs will be equal to the sum of the number of ways for the previous two numbers of stairs?

If you answered YES, why does the explanation for 6 stairs convince you for any number of stairs? If you answered NO, what would you need, in order to be convinced?

Of 17 students, the (anonymous) responses of 15 indicated that the particular example - the explanation for 6 stairs - was, for them, generic in relation to other numbers of stairs. The following responses are typical:

> [Student A] If you start with a one, you have 14 left. If you start with a two, you have 13 left. So the sum of these two will form the same formula as for 6 stairs.

> [Student B] We could re-write the explanation in terms of n, $n-1$ and $n-2$ where $n = 6$ stairs in the explanation and $5 = n-1$ and $4 = n-2$. So we see we had the correct method in the explanation above.

Of the two who were unconvinced (questions 2 and 3), one was explicit about lack of conviction about the initial particular explanation (for 6 stairs), being slightly unsure that it accounted for all possibilities. The other seemed to require further confirming instances of the generalisation before s/he could "accept it".

A group of teachers who had joined a study group to work on teaching A-level mathematics was set a similar problem:

> In how many ways can n 1 by 2 rectangles be arranged to form a 2 by n rectangle?

Three of the teachers in the group had each reached the conclusion that the numbers of ways of forming the rectangles were the Fibonacci numbers. Having been encouraged to think about why this was the case, one of the group gave a demonstration to the two others that the 4 by 2 rectangles could all be formed by adding a single 1 by 2 rectangle to a 3 by 2 rectangle, or by adding two 1 by 2 rectangles to a 2 by 2 rectangle.

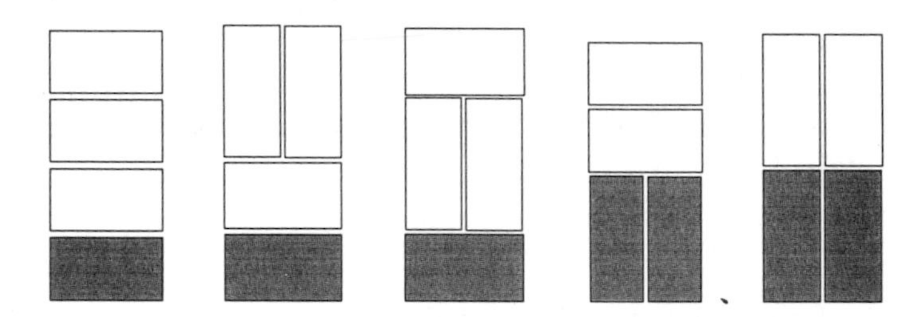

An observer might have interpreted his demonstration generically and assumed that he and his two listeners would do likewise. However after giving his demonstration he said: "Now - I haven't tried, but I guess three and four - I'm just assuming at the moment that it's just adding on ..." and he proceeded to form the 5 by 2 rectangles by adding on to the 4 by 2 and 3 by 2 rectangles. Even though he was aware of the generality which this particular demonstration pointed to, the demonstration in the single case was not enough to satisfy him. He wanted to consider another particular example before he would convince himself that the Fibonacci sequence of numbers of arrangements was justified.

Theorems and generic proofs

The 'Stairs' and 'Rectangles' examples are open to criticism on the grounds that, in common with most 'investigations' designed for the classroom, they are concerned with 'content' on the margins of mainstream school/college mathematics. Partly with this complaint in mind, a similar study was carried out with a class of second-year students on the Mathematics/Education course referred to above. The context was a 48-hour unit in the Theory of Numbers, covering the usual undergraduate topics found in texts such as Baker (1984) or Davenport (1992). Building on the earlier 'processes' module mentioned above, the lecturer for this unit regularly exploits generic examples as a forerunner of (sometimes instead of) conventional general algebraic arguments. Modular arithmetic arises early in the course, with the theorem that every prime number p has a 'primitive root' (that is to say, the group $\{1, 2, 3, ..., p-1\}$ under multiplication modulo p is cyclic). The standard general proof (see, for example, Baker, 1984, p. 23) is surprisingly indirect and overburdened with notational complexity. This latter factor is, in part, due to the fact that it has to deal with a double layer of generality: any prime p and any divisor d of $p-1$. The lecturer's proof focused on the case $p = 19$ and $d = 6$, with the structure of the generic exposition summarised afterwards as follows below. The summary presented here omits many details which are not essential to our argument concerning generic proof. Step 2 uses the fact that p is prime in establishing that if $N_6 \neq 0$, then $N_6 = \phi(6)$.

The prime number $p = 19$ has a primitive root

1. The group M_{19} has 18 elements, so the order of each of those elements must divide 18. Possibilities are 1, 2, 3, 6, 9 and 18.
   Suppose there are N_1 elements of order 1, N_2 of order 2, ..., N_{18} of order 18, so $N_1 + N_2 + N_3 + N_6 + N_9 + N_{18} = 18$. To prove that 19 has a primitive root, **we need to demonstrate that $N_{18} \neq 0$.**
2. Focus for the moment on the elements of order 6 (there might be none). Argue (as in the lecture) that if $N_6 \neq 0$, then $N_6 = \phi(6)$. A similar line of reasoning would establish that $N_1 = 0$ or $N_1 = \phi(1)$, $N_2 = 0$ or $N_2 = \phi(2)$, ..., $N_{18} = 0$ or $N_{18} = \phi(18)$. It follows that $[\phi(i) - N_i] = 0$ for each of the six possible values of i.
3. We know that $\sum_{d|n} \phi(d) = n$ and so $\phi(18) = \phi(1) + \phi(2) + \phi(3) + \phi(6) + \phi(9) + \phi(18) = 18$.
   Since $N_1 + N_2 + N_3 + N_6 + N_9 + N_{18} = 18$, it follows that
   $[\phi(1) - N_1] + [\phi(2) - N_2] + [\phi(3) - N_3] + [\phi(6) - N_6] + [\phi(9) - N_9] + [\phi(18) - N_{18}] = 0$.
   Since each bracket is non-negative, they must all be zero.
4. In other words, $\phi(1) = N_1$, $\phi(2) = N_2$, ... and in particular, $\boldsymbol{\phi(18) = N_{18}}$.
   Now $\phi(18) = 6$, so it follows that **19 has a primitive root** - 6 of them, in fact.

The students were asked to make a written response to three questions (analogous to those asked in the previous 'Stairs' study):

1. Are you happy with the above explanation with $p = 19$ i.e. is it convincing?

2. Does the above explanation help to convince you that 29 has a primitive root?

3. Does the first explanation [for the case $p = 19$] convince you that *every* prime has a primitive root?

As with the 'Stairs' study, these students were asked to elaborate on their responses. The 19 (by coincidence) returns indicate that the argument concerning the case $p = 19$ was generic for 12 of the students, whose responses included the following:

> [concerning $p = 29$] The whole process could be repeated using 29 so that 1, 2, 4, 7, 14, 28 are possible orders e.g. 7, same argument as for M_{19}. Come to same conclusions, you just have different numbers involved.
>
> I went through the proof with $p = 29$ and felt that it was applicable.
>
> You can adapt the proof so that it would apply to the possible orders of M_{29}, and then follow through the same argument.
>
> M_{29} has 28 elements with divisors 1, 2, 4, 7, 14, 28 [...] $N_1 + N_2 + \dots N_{28} = 28$, similarly $\phi(28) = 28$ because of the theorem [...] I can see that the argument can be transferred to $p = 29$ and would show that $N_{28} = \phi(28) = 12$. Therefore 29 has 12 primitive roots. Quite convincingly!!!
>
> [concerning any prime] It is easy to follow the logical progression of the proof for $p = 19$ with any other prime in mind, and I can see no area of the proof which gives me any doubt that it wouldn't work for any prime.

Some respondents stated that they appreciated the 'concreteness' of the generic argument:

> If it was general with no numbers, I think I would get confused.

Two students, moreover, volunteered that the generic argument with $p = 19$ indicated (for them) how they might formulate a general one:

> because you could just extend the argument for any prime and substitute in the values, and perhaps produce a general form for the proof.
>
> By changing 19 to p, I could generalise the argument.

For four of the students, the generic intention of the example was not effective. They were united by the sense that the proof for $p = 19$ was precisely that, and no more:

> Although the explanation for $p = 19$ is clear and true, it doesn't necessarily follow that $p = 29$ has a primitive root. So I'd prefer to work through it [$p = 29$] before I was convinced.

In effect, these students are being cautious about what they perceive as a case of empirical generalisation. They fail to see that the argument has been presented with the aim of *structural* generalisation; rather, the argument is not effective in suggesting structural generalisation to them. It could be, of course, that these second-year students are conditioned into being satisfied with nothing less than a 'general' algebraic argument: indeed, they all indicated their need for a "general proof" before they could be convinced. One student spoke for all four when s/he wrote:

> I would like a general proof to reassure me for all primes.

although s/he added somewhat wistfully:

> however, a general proof on its own would probably confuse me.

The ambivalence of the three remaining responses could not be resolved, since all the returns were anonymous; on the whole, these respondents found the case for primes other than 19 to be plausible but not entirely convincing. The distinction between public accountability and personal conviction is brought out in the following:

> It needs to be proved generally [...] but using a number 19 makes it easier to follow [...] A general proof needs to be followed through step by step. (But I am fairly convinced already - I can't see why it wouldn't work.)

As Walther (*ibid.*) indicates, the psychological effectiveness of the generic proof hinges on the identification and transfer of structural invariants, as opposed to contingent variables, in the argument. The sophisticated mathematician is able to isolate such invariants; indeed, some of the quotations from the students' writing above bring this out explicitly, such as: "You can *adapt* the proof"; "the argument can be *transferred*". One student uses the word "*similarly*" to refer to this transfer from one particular case to another, and we return to this matter in a moment.

As Krutetskii (1976) has observed:

> Capable pupils [...] very easily find the essential and general in the particular ... (p. 262)

The same kind of sophistication allows a mathematician to see that the working produced below to prove that a particular transformation T maps the line $y = 2x + 1$ to another straight line is "essentially the same" as that required to prove that any affine transformation maps any straight line to another straight line.

I am going to define a transformation T as follows:

$$T: \begin{pmatrix} x \\ y \end{pmatrix} \rightarrow \begin{pmatrix} 2 & 3 \\ 1 & 2 \end{pmatrix} \begin{pmatrix} x \\ y \end{pmatrix} + \begin{pmatrix} 4 \\ 5 \end{pmatrix}.$$

I want to look at the effect of this transformation on a particular set of points, namely the line $y = 2x + 1$. I'm interested in whether the image set is a particular curve - that is whether there is an algebraic relationship between the x and y co-ordinates of any point in the image set. If I call these x and y co-ordinates X and Y, then:

$$X = 2x + 3y + 4$$
$$X = x + 2y + 5$$

But if the point (x, y) lies on the line $y = 2x + 1$ then

$$X = 2x + 3(2x + 1) + 4 = 8x + 7$$
$$Y = x + 2(2x + 1) + 5 = 5x + 7$$

So $$Y = \frac{5}{8}(8x + 7) - \frac{5}{8} \text{ x } 7 + 7$$
$$= \frac{5}{8}(8x + 7) + \frac{21}{8}$$
$$= \frac{5}{8}X + \frac{21}{8}$$

I have found a linear relationship between X and Y and so I know that the image of the line $y = 2x + 1$ under the transformation T is another straight line.

We suggest that learners of mathematics, at all levels, can be assisted to perceive and value that which is generic in particular insights, explanations and arguments - especially their own. The barrier between such a level of knowing and the writing of 'general' proofs is then seen for what it is - a lack of fluency not with ideas, but with notation.

'Similarly'

Semadeni (1984) highlights a didactic issue concerning the acceptability of generic ['action'] proofs:

> an action proof [...] involves a psychological question: how can one know whether the child is convinced of the validity of the proof by inner understanding and not just by being prompted by the authority of the teacher? Without dismissing this criticism, we note that it applies to any

> proof in a textbook: if the author finds his proof correct and complete, this does not automatically imply that students understand it. (p. 34)

As we have shown, some of the students in the 'primitive roots' study were not prepared to accept a generic argument as a substitute for a 'general' proof. In this, they reflect a prevailing (if relatively recent) attitude to what is required in the formulation of a 'proper' proof, even though the general form "might confuse" an inexperienced audience more than a generic argument. At the heart of this conventional dismissal of the generic proof as a substitute for the general (in lecture notes and certainly in textbooks), there appears to lie a greater concern by teachers and authors to meet the formalist requirements of 'mathematics' than to meet the psychological requirements of the audience. In fact, a conventionally acceptable proof may make some appeal to "the identification and *transfer* of structural invariants" by subtle use of the adverb "similarly", in order to achieve economy of exposition. Consider, for example, the following standard proof that equivalence classes are disjoint or equal:

> Let R be an equivalence relation on a set X. For $a \in X$, denote the equivalence class $\{x \in X\colon aRx\}$ by $R(a)$. Suppose $R(a) \cap R(b)$ is not empty, and contains y. Let $x \in R(a)$. Then aRx; also aRy and bRy. By the symmetry of R, yRa. Hence bRa by transitivity. Finally, bRx by transitivity, and we conclude that $x \in R(b)$. It follows that $R(a) \subseteq R(b)$. **Similarly**, $R(b) \subseteq R(a)$, and thus $R(a) = R(b)$.

The transfer which is required here is by symmetry rather than by analogy. A portion of the argument - running from the fourth sentence as far as the penultimate one - remains valid if a and b are transposed. The invariants of the argument are the common membership of the intermediary element y, the role of x as a representative of one of the equivalence classes, and the defining characteristics of R which support the detail of the symbolic chase to membership of the other equivalence class. Although the mature reader may readily discern the role of each of the variables, this is not a trivial matter. Our point is that the proof is formulated in a way that demands some consent, some co-operation from the reader. The generic proof demands something similar, except that the transfer required is by way of adaptation or substitution.

Summary

We recognise that inductive reasoning of a quasi-empirical character motivates students, and lends an authentic air of discovery to the mathematics classroom. Such activity transcends empirical speculation when explanation is available to the student as a structural generalisation of some kind. A generic example which successfully "speaks the generality" (Mason and Pimm, 1984, p. 284) for the audience has the quality of such a structural generalisation.

References

Baker, A.: 1984, *A Concise Introduction to the Theory of Numbers*, Cambridge University Press, Cambridge.

Balacheff, N.: 1988, 'Aspects of proof in pupils' practice of school mathematics', in Pimm, D. (ed.), *Mathematics, Teachers and Children*, Hodder and Stoughton, London, pp. 216-235.

Burn, R. P.: 1982, *A Pathway into Number Theory*, Cambridge University Press, Cambridge.

Davenport, H.: 1992, *The Higher Arithmetic* (6th edition), Cambridge University Press, Cambridge.

Krutetskii, V. A.: 1976, *The Psychology of Mathematical Abilities in Schoolchildren,* University of Chicago Press, Chicago.

Laplace, P. S.: 1901, *A Philosophical Essay on Probabilities*, Truscott and Emory, New York.

MacLane, S.: 1986, *Mathematics: Form and Function*, Springer-Verlag, New York.

Mason, J. and Pimm, D.: 1984, 'Generic examples: seeing the general in the particular', *Educational Studies in Mathematics*, **15**, 277-289.

Polya, G.: 1954, *Mathematics and Plausible Reasoning, Volume 1: Induction and Analogy in Mathematics*, Princeton University Press, Princeton, NJ.

Rescher, N.: 1980, *Induction: an essay on the justification of inductive reasoning*, Basil Blackwell, Oxford.

Semadeni Z.: 1984, 'Action proofs in primary mathematics teaching and in teacher training', *For the Learning of Mathematics*, **4**(1), 32-34.

Stamp, M.: undated, circa 1980, 'Perimeter equals area', in *Topics in Mathematics*, Association of Teachers of Mathematics, Derby, p. 3.

Walther G.: 1984, 'Action proofs vs. illuminating examples?', *For the Learning of Mathematics*, **4**(3), 10-12.